CAMBRIDGE LIBRARY COLLECTION

Books of enduring scholarly value

Travel and Exploration

The history of travel writing dates back to the Bible, Caesar, the Vikings and the Crusaders, and its many themes include war, trade, science and recreation. Explorers from Columbus to Cook charted lands not previously visited by Western travellers, and were followed by merchants, missionaries, and colonists, who wrote accounts of their experiences. The development of steam power in the nineteenth century provided opportunities for increasing numbers of 'ordinary' people to travel further, more economically, and more safely, and resulted in great enthusiasm for travel writing among the reading public. Works included in this series range from first-hand descriptions of previously unrecorded places, to literary accounts of the strange habits of foreigners, to examples of the burgeoning numbers of guidebooks produced to satisfy the needs of a new kind of traveller - the tourist.

Tractatus de Globis et Eorum Usu

The publications of the Hakluyt Society (founded in 1846) made available edited (and sometimes translated) early accounts of exploration. The first series, which ran from 1847 to 1899, consists of 100 books containing published or previously unpublished works by authors from Christopher Columbus to Sir Francis Drake, and covering voyages to the New World, to China and Japan, to Russia and to Africa and India. Robert Hues (1553–1632) was an English mathematician and geographer who published this work in 1594 to explain the use of the new terrestrial and celestial globes devised by Emery Molyneux in 1592. These were the first English manufactured globes and were popular with both navigators and students. The five parts of this book describe these globes and explain their use in calculating fundamental navigational points, providing valuable insights into their appearance and practical application in early sixteenth-century navigation.

Cambridge University Press has long been a pioneer in the reissuing of out-of-print titles from its own backlist, producing digital reprints of books that are still sought after by scholars and students but could not be reprinted economically using traditional technology. The Cambridge Library Collection extends this activity to a wider range of books which are still of importance to researchers and professionals, either for the source material they contain, or as landmarks in the history of their academic discipline.

Drawing from the world-renowned collections in the Cambridge University Library, and guided by the advice of experts in each subject area, Cambridge University Press is using state-of-the-art scanning machines in its own Printing House to capture the content of each book selected for inclusion. The files are processed to give a consistently clear, crisp image, and the books finished to the high quality standard for which the Press is recognised around the world. The latest print-on-demand technology ensures that the books will remain available indefinitely, and that orders for single or multiple copies can quickly be supplied.

The Cambridge Library Collection will bring back to life books of enduring scholarly value (including out-of-copyright works originally issued by other publishers) across a wide range of disciplines in the humanities and social sciences and in science and technology.

Tractatus de Globis et Eorum Usu

A Treatise Descriptive of the Globes Constructed by Emery Molyneux and Published in 1592

Robert Hues
Edited by Clements R. Markham

CAMBRIDGE UNIVERSITY PRESS

Cambridge, New York, Melbourne, Madrid, Cape Town, Singapore,
São Paolo, Delhi, Dubai, Tokyo, Mexico City

Published in the United States of America by Cambridge University Press, New York

www.cambridge.org
Information on this title: www.cambridge.org/9781108013499

This edition first published 1889
This digitally printed version 2010

ISBN 978-1-108-01349-9 Paperback

WORKS ISSUED BY

The Hakluyt Society.

ROBERT HUES'

TRACTATUS DE GLOBIS.

SAILING DIRECTIONS

FOR THE

CIRCUMNAVIGATION OF ENGLAND.

No. LXXIX.

THE MOLYNEUX CELESTIAL GLOBE.

One of a Pair at the Middle Temple Library.

(AFTER A PHOTOGRAPH.)

TRACTATUS DE GLOBIS

ET EORUM USU.

A TREATISE

DESCRIPTIVE OF THE GLOBES CONSTRUCTED BY EMERY MOLYNEUX, AND PUBLISHED IN 1592.

BY

ROBERT HUES.

Edited, with Annotated Indices and an Introduction,

BY

CLEMENTS R. MARKHAM, C.B., F.R.S.

LONDON:
PRINTED FOR THE HAKLUYT SOCIETY,
4, LINCOLN'S INN FIELDS, W.C.

M.DCCC.LXXXIX.

LONDON:
WHITING AND COMPANY, SARDINIA STREET, LINCOLN'S INN FIELDS.

COUNCIL

OF

THE HAKLUYT SOCIETY.

CONTENTS.

Fifth Part.

ILLUSTRATION.

INTRODUCTION.

At the time when English sailors began to make the reign of the great Queen illustrious by daring voyages and famous discoveries, it was natural that these deeds should be worthily recorded. When Drake and Cavendish had circumnavigated the globe, when Raleigh had planted Virginia, Davis had discovered his Straits, and Lancaster had found his way to India, the time had come for Hakluyt to publish his *Principal Navigations*, and for Molyneux to construct his Globes.

Englishmen were coming to the front rank as discoverers and explorers, and it naturally followed that maps and globes should be prepared by their countrymen at home, which should alike record the work already achieved and be useful for the guidance of future navigators. But the construction of globes entailed considerable expense, and there was need for liberal patronage to enable scientific men to enter upon such undertakings.

In the days of Queen Elizabeth the merchants of England were ever ready to encourage enterprises having for their objects the improvement of navigation and the advancement of the prosperity of their country. While the constructor of the first globes ever made in this country received help and advice from navigators and mathematicians, he was liberally supplied with funds by one of the most munificent of our merchant princes. The appearance of the globes naturally created a great sensation, and much interest was taken in appliances which were equally useful to the student and to the practical navigator. Two treatises intended to describe these new appliances, and to serve as guides for their use, were published very soon after their completion. One of these, the *Tractatus de Globis* of the celebrated mathematician, Robert Hues, has been selected for republication by the Hakluyt Society. Before describing the Molyneux Globes, and the contents of the Guide to their use, it will be well to pass in review the celestial and terrestrial globes which preceded, or were contemporaneous with, the first that was made in England, so far as a knowledge of them has come down to us.

The celestial preceded the terrestrial globes by many centuries. The ancients appear to have adopted this method of representing the heavenly bodies and their movements at a very early period. Diodorus Siculus asserts that the use of the globe was first discovered by Atlas of Libya, whence originated the fable of his bearing up the heavens on his shoulders. Others attribute the invention to Thales; and subsequent geographers, such as Archimedes, Crates, and Proclus, are said to have improved upon it. Posidonius, who flourished 150 B.C., and is often quoted by Strabo, constructed a revolving sphere to exhibit the motions of the heavenly bodies; and three hundred years afterwards Ptolemy laid down rules for the construction of globes. There are some other allusions to the use of globes among ancient writers; the last being contained in a passage of Leontius Mechanicus, who flourished in the time of Justinian. He constructed a celestial globe in accordance with the rules of Ptolemy, and after the description of stars and constellations given by Aratus. Globes frequently occur on Roman coins. Generally the globe is merely used to denote universal dominion. But in some instances, especially on a well-known medallion of the Emperor Commodus, a celestial globe, copied, no doubt, from those in use at the time, is clearly represented. No Greek or Roman globes have, however, come down to us. The oldest in existence are those made by the Arabian astronomers.

The earliest form appears to have been the armillary sphere, consisting of metal rings fixed round a centre, and crossing each other on various planes, intended to represent the orbits of heavenly bodies. The Arab globes were of metal, and had the constellations and fixed stars engraved upon them. At least five dating from the thirteenth century have been preserved. One is in the National Museum at Naples, with the date 1225. Another, dated 1275, belongs to the Asiatic Society of London; and a third, dated 1289, is at Dresden. There are two others, without date, but probably to be referred to the same period, one belonging to the

Astronomical Society of London, the other to the National Library of Paris.

But the most ancient celestial globe is at Florence, and has been described by Professor Meucci.[1] It belongs to the eleventh century.

The astronomical knowledge of the Arabs in the East was communicated to their countrymen in Spain, and the schools of Cordova became so famous that they were frequented by students from Christian Europe; among whom was the celebrated mathematician, Gerbert d'Auvergne, afterwards Pope Silvester II. Valencia was one of the most flourishing centres of Arabian culture in Spain, at first under the Khâlifahs of Cordova, and from 1031 to 1094 as the capital of a small, independent kingdom. It was in Valencia that the celestial globe, now at Florence, was constructed, in the year 1070 A.D.[2] It is 7.8 inches in diameter. All the forty-seven constellations of Ptolemy are engraved upon

[1] *Il Globo Celeste Arabico del Secolo XI esistente nel gabineto degli strumenti antichi di astronomia, di fisica, e di matematica del R. Instituto di Studi Superiori illustrato da F. Meucci* (Firenze, 1878).

[2] Professor Meucci observed that the star *Regulus* was placed on the globe at a distance of 16° 40′ from the sign of Leo. Ptolemy, in 140 A.D., gave this distance as 2° 30′. According to Albategnius, the star advances 1° in sixty-six years. It had moved 14° 10′ since 140 A.D., which would give 1070 as about the date of the globe.

The Arabic inscription on the globe coincides remarkably with this calculation. It states that the globe was made at Valencia by Ibrahim ibn Said-as-Sahli, and his son Muhammad, in the year 473 of the Hegira, equivalent to 1080 A.D. It was conit, except the "*Cup*", and 1,015 stars are shown, with the different magnitudes well indicated. It is a very precious relic of the civilisation of the Spanish Arabs, and is specially interesting as the oldest globe in existence, and as showing the care with which the Arabian astronomers preserved and handed down to posterity the system of Ptolemy. The globe possessed by the Emperor Frederick II, with pearls to indicate the stars, doubtless resembled those of the same period which have come down to us.

The oldest terrestrial globe in existence is that constructed by Martin Behaim, at Nuremburg, in 1492. It is made of pasteboard covered with parchment, and is 21 inches in diameter. The only lines drawn upon it are the equator, tropics, and polar circles, and the first meridian, which passes through Madeira. The meridian is of iron, and a brass horizon was added in 1500. The globe is illuminated and ornamented, and is rich in legends of interest and in geographical details. The author of this famous globe was born at Nuremburg of a good family. He had studied under Regiomontanus. He settled and married at Horta, the capital of Fayal, in the Azores, had made numerous voyages, and had been in the exploring expedition with Diogo Cam when that Portuguese navigator discovered the mouth of the Congo. Behaim had the reputation of being a good astronomer, and is said by

structed for Abû Isa ibn Labbun, a personage of note in the political and literary history of Muslim Spain during that century.

Barros[1] to have invented a practical instrument for taking the altitude of the sun at sea.

Baron Nordenskiöld considers that the globe of Behaim is, without comparison, the most important geographical document that saw the light since the atlas of Ptolemy had been produced in about 150 A.D. He points out that it is the first which unreservedly adopts the existence of antipodes, the first which clearly shows that there is a passage from Europe to India, the first which attempts to deal with the discoveries of Marco Polo. It is an exact representation of geographical knowledge immediately previous to the first voyage of Columbus.

The terrestrial globe next in antiquity to that of Behaim is dated 1493. It was found in a shop at Laon, in 1860, by M. Léon Leroux, of the Administration de la Marine at Paris. It is of copper-gilt, engraved, with a first meridian passing through Madeira, meridian-lines on the northern hemisphere at every fifteen degrees, crossed by parallels corresponding to the seven climates of Ptolemy. There are no lines on the southern hemisphere. The author is unknown, but M. D'Avezac considered that this globe represented geographical knowledge current at Lisbon in about 1486. It appears to have been part of an astronomical clock, or of an armillary sphere, for it is only 6½ inches in diameter.[2]

Baron Nordenskiöld was the first to point out

[1] *Dec. I*, lib. iv, cap. 2.

[2] D'Avezac gives a projection of the Laon globe in the *Bulletin de la Société de Géographie de Paris*, 4me Série, viii (1860).

that a globe constructed by John Cabot is mentioned in a letter from Raimondo di Soncino to the Duke of Milan, dated December 18th, 1497. But it does not now exist.

The earliest post-Columbian globe in existence dates from about A.D. 1510 or 1512. It was bought in Paris by Mr. R. M. Hunt, the architect, in 1855, and was presented by him to Mr. Lenox of New York; it is now in the Lenox Library. This globe is a spherical copper box 4½ inches in diameter, and is pierced for an axis. It opens on the line of the equator, and may have been used as a *ciborium*. The outline of land and the names are engraved on it, but there is no graduation. The author is unknown.

Among the papers of Leonardo da Vinci at Windsor Castle there is a map of the world drawn on eight gores, which appears to have been intended for a globe. It is interesting as one of the first maps on which the name America appears. Mr. Major has fully described this map in a paper in the *Archæologia*,[1] and he believes that it was actually drawn by Leonardo da Vinci himself. But Baron Nordenskiöld gives reasons for the conclusion that it was copied from some earlier globe by an ignorant though careful draughtsman.

In 1881 some ancient gores were brought to

[1] "A Memoir on a Mappemonde by Leonardo da Vinci, being the earliest map hitherto known containing the name of America; now in the royal collection at Windsor." By R. H. Major, Esq., F.S.A. (*Archæologia*, vol. xl, 1865).

light by M. Tross, in a copy of the *Cosmographiæ Introductio* of Waldseemüller, printed at Lyons in 1514 or 1518. They are from engravings on copper by Ludovicus Boulenger.

A globe was constructed at Bamberg in 1520, by Johann Schoner of Carlstadt, which is now in the town library at Nuremburg; it consists of twelve gores. There is a copy of the Schoner globe, 10½ inches in diameter, at Frankfort,[1] and two others in the Military Library at Weimar. On the Schoner globe, North America is broken up into islands, but South America is shown as a continuous coast-line, with the word America written along it, as on the gores attributed to Leonardo da Vinci.[2] Florida appears on it, and the Moluccas are in their true positions. A line shows the track of Magellan's ships; and the globe may be looked upon as illustrating the history of the first circumnavigation.

A beautiful globe was presented to the church at Nancy by Charles V, Duke of Lorraine, where it was used as a *ciborium*. It is now in the Nancy public library. It is of chased silver-gilt and blue enamel, 6 inches in diameter.[3]

[1] The Frankfort globe is given by Jomard in his *Monuments de la Géographie*; see also *J. R. G. S.*, xviii, 45.

[2] *Johann Schoner, Professor of Mathematics at Nuremburg. A reproduction of his Globe of* 1523, *long lost.* By Henry Stevens of Vermont; edited, with an Introduction and Bibliography, by C. H. Coote (London, 1888).

[3] First described by M. Blau, *Mémoires de la Société Royale de Nancy*, 1825, p. 97.

There is a globe in the National Library at Paris very like that of Schoner, which has been believed to be of Spanish origin. Another globe in the same library, with the place of manufacture—"Rhotomagi" (Rouen)—marked upon it, but no date, is supposed to have been made in 1540. It belonged to Canon L'Ecuy of Premontré. This globe was the first to show North America disconnected with Asia.

In 1541 Gerard Mercator completed his terrestrial globe at Louvain, dedicating it to Cardinal Granvelle. Its celestial companion was finished ten years afterwards. These globes were 16 inches in diameter. Many replicas were produced, and Blundeville[1] alludes to them as in common use in England in 1594. Yet only two sets now exist. In May 1868 the twelve gores for one of these was bought by the Royal Library of Brussels, at the sale of M. Benoni-Verelst of Ghent. The other was found in 1875 at the Imperial Court Library of Vienna. The terrestrial globe has rhumb lines, which had hitherto only been shown on plane-charts. The celestial globe has fifty-one constellations, containing 934 fixed stars.

1 Thomas Blundeville was a country gentleman, born in 1568. He succeeded to Newton Flotman, in Norfolk, in 1571; and was an enthusiastic student of astronomy and navigation. In 1589 he published his *Description of universal mappes and cardes*, and his *Exercises* appeared in 1594. This work was very popular among the navigators of the period, and went through at least seven editions. Blundeville also wrote on horsemanship. His only son was slain in the Low Countries.

A copper globe was constructed at Rome by Euphrosinus Ulpius in 1542, and dedicated to Pope Marcellus II when he was a cardinal. It was bought in Spain in 1859, and is now in the library of the New York Historical Society. It is 15½ inches in diameter, divided in the line of the equator, and fastened by iron pins, and it has an iron cross on the North Pole. Its height, with the stand, is 3 feet 8 inches. The meridian-lines are at distances of 30°, the first one passing through the Canaries. Prominence is also given to the line of demarcation between Spain and Portugal, laid down by Pope Alexander VI. There is another globe, found at Grenoble in 1855, and now in the National Library at Paris, by A. F. von Langeren, which may possibly antedate the Molyneux globes.[1]

In the Oldnorske Museum at Copenhagen there is a small globe of 1543, mounted as an armillary sphere, with eleven brass rings. It was constructed by Caspar Vopell, and is believed to have belonged to Tycho Brahe. A small silver globe is part of the Swedish regalia, and was made in 1561 for the coronation of Eric XIV. Similar globes, forming goblets or *ciboires*, are preserved in the Rosenborg Palace at Copenhagen and in the Museum at Stockholm. They are merely specimens of goldsmiths'

[1] After the globes of Molyneux followed those of Blaew and Hondius. Langeren and Hondius were rivals. They announced their intention of bringing out two globes in 1597, but no copies are known to exist. The globes of W. Janssen Blaew (1571-1638) were of wood, the largest being 27 inches in diameter, the smallest 7½ inches.

work, useful only if other maps of the same period were wanting.

Counting the gores of Tross and of Leonardo da Vinci, there are thus twelve terrestrial globes now in existence which preceded the first that was constructed in England.

The preparation of celestial globes and armillary spheres received an impetus from the labours of the great astronomers who flourished for two centuries, from the time of Copernicus to that of Galileo.

Nicolaus Copernicus was born at Thorn on the Vistula in 1473, and was educated at the University of Cracow, studying medicine and painting, as well as mathematics. After passing some years at the University of Bologna and at Rome, he returned to his native country. The uncle of Copernicus was Bishop of Warmia or Warmland, on the Baltic, near Danzig; with a cathedral at Frauenburg, on the shores of the Friske Haff. Here the great astronomer became a canon; here he passed the remainder of his life; and here he wrote his great work, *De Revolutionibus Orbium Cœlestium.* It was completed in 1530, but over ten more years were devoted to the work of correcting and altering, and when, at last, it was printed at Nuremberg, Copernicus was on his death-bed. He died on May 23rd, 1543, having just lived long enough to rest his hand on a printed copy of his work. It is not known that a sphere was ever constructed in his lifetime to illustrate his system. Tycho Brahe was born at Knudstrup, in December 1546, three years after the death of Copernicus. The one was a quiet ecclesiastic; the other a man of noble birth, whose career was surrounded by difficulties, owing to the family prejudices, which were irreconcilable with the studies and occupations of his choice. The family of Tycho Brahe believed that the career of arms was the only one suited for a gentleman. He became a student at Copenhagen and at Wittenberg, and still further offended his relations by marrying a beautiful peasant girl of Knudstrup. The accident of his birth made it impossible for him to avoid strife. At Rostock he felt bound to fight a duel with a Dane named Pasberg, to decide the question as to which was the best mathematician. Tycho Brahe had half his nose cut off, and ever afterwards he wore a golden nose. But, in spite of obstacles, he rose to eminence as an astronomer. He discovered errors in the Alphonsine Tables, and observed a new star in Cassiopeia in 1572. King Frederick II of Denmark recognised the great merits of Tycho Brahe. He granted him the island of Hveen in 1576, where the illustrious astronomer built his château of Uranienberg and his observatories.[1] Here he made his catalogue of stars, and here he lived and observed for many years; but, on the death of Frederick in 1588, the enemies of the great man poisoned the mind of Christian IV against him. His pension and all his allowances were withdrawn,

[1] The instruments of Tycho Brahe and a plan of Uranienberg are given in vol. i of the *Atlas Major* of Blaew (Blasius).

and he was nearly ruined. In 1597 he left the island, and set sail, with his wife and children, for Holstein. In 1599 he accepted a cordial invitation from the Emperor Rudolph II to come to Bohemia, and was established in the Castle of Beneteck, five miles from Prague. He died at Prague in 1601, aged 55.

The celestial globe constructed by Tycho Brahe is described by his pupil Pontanus. It was made of wood covered with plates of copper, and was six feet in diameter. It was considered to be a magnificent piece of work, and many strangers came to the island of Hveen on purpose to see it. But when Tycho Brahe was obliged to leave Denmark, he took the globe with him, and it was eventually deposited in the imperial castle at Prague. Of about the same date is the celestial globe at the South Kensington Museum, made for the Emperor Rudolph II at Augsburg in 1584. It is of copper-gilt, and is 7½ inches in diameter.

John Kepler, who was born at Weil in Würtemberg in 1571, is also said to have been of noble parentage; but his father was so poor that he was obliged to keep a public-house. A weak and sickly child, Kepler became a student at Tübingen, and devoted himself to astronomical studies. He visited Tycho Brahe at Prague in 1600, and succeeded him as principal mathematician to the Emperor Rudolph II. But he was always in pecuniary difficulties, and was irritable and quick-tempered, owing to ill-health and poverty. Nevertheless, he made great advances in the science of astronomy. He completed the Rudolphine Tables in 1627, being the first calculated on the supposition that the planets move in elliptical orbits. Kepler's laws relate to the elliptic form of orbits, the equable description of areas, and to the proposition that the squares of the periodic times are proportional to the cubes of the mean distances from the sun. His work on the motions of the planet Mars was published in 1609. Kepler died in November 1630, aged 58.

The great Italian astronomer was his contemporary. Galileo Galilei was born at Pisa in 1564, and was educated at the university of his native town. Here he discovered the isochronism of the vibrations of the pendulum; and in 1592, when professor at Padua, he became a convert to the doctrines of Copernicus. His telescope, completed in 1609, enabled him to discover the ring of Saturn and the satellites of Jupiter; while the latter discovery revealed another method of finding the longitude. The latter years of the life of Galileo were clouded by persecution and misfortune. The Convent of Minerva at Rome, where stupid bigots forced him to recant, and where he whispered "e pur se muove", is now the Ministry of Public Instruction of an enlightened government. His trial before the Inquisition was in 1632; he lost his daughter in 1634; and in 1636 he became blind. Galileo died in the arms of his pupil Viviani, in January 1642. There can be no more fitting monument to the great astronomer than the "Tribuna" which has been erected to his honour at Florence. Frescoes of the chief events in his life adorn the walls, while his instruments, and those of his pupils Viviani and Torricelli, illustrate his labours and successes.

Pontanus, who was a disciple of Tycho Brahe, mentions that Ferdinand I of Tuscany had two large globes, one terrestrial, and the other an armillary sphere with circles and orbs, both existing in the time of Galileo. The latter, which was designed by the cosmographer Antonio Santucci between 1588 and 1593, is still preserved, and has been described by Professor Meucci.[1] It is constructed on the Ptolemaic system, and consists of nine concentric spheres, the outer one being 7 feet in diameter, and the earth being in the centre. The frame rests on a pedestal consisting of four caryatides, which represent the four cardinal points; and it stands near the entrance to the "Tribuna" of Galileo. It is the last and most sumptuous illustration of the old Ptolemaic system, and a monument of the skill and ingenuity of the scientific artists of Florence.

The celestial globe of Tycho Brahe and the armillary sphere of Santucci cannot have been seen by Molyneux. Their construction was nearly contemporaneous with that of the first English globes. But all the

[1] *La Sfera Armillare di Tolomeo, construita da Antonio Santucci* (Firenze, 1876).

other globes that have been enumerated preceded the kindred work of our own countrymen ; and in their more complete development, under the able hands of Mercator, they served as the pattern on which our mathematician built up his own enlarged and improved globes.

We find very little recorded of Emery Molyneux, beyond the fact that he was a mathematician residing in Lambeth. He was known to Sir Walter Raleigh, to Hakluyt, and to Edward Wright, and was a friend of John Davis the Navigator. The words of one of the legends on his globe give some reason for the belief that Molyneux accompanied Cavendish in his voyage round the world. The construction of the globes appears to have been suggested by learned men to Mr. William Sanderson, one of the most munificent and patriotic of the merchant-princes of London, in the days of the great Queen. He fitted out the Arctic expeditions of Davis ; and the same liberal patron readily undertook to defray the expenses connected with the construction of the globes. There are grounds for thinking that it was Davis who suggested to Mr. Sanderson the employment of Emery Molyneux. The approaching publication of the globes was announced at the end of the preface to the first edition of Hakluyt's *Voyages*, which saw the light in 1589. There was some delay before they were quite completed, but they were actually published in the end of 1592.

The Molyneux globes are 2 feet 2 inches in

diameter,[1] and are fixed on stands. They have graduated brass meridians, and on that of the terrestrial globe a dial circle or "Horarius" is fixed. The broad wooden equator, forming the upper part of the stand, is painted with the zodiac signs, the months, the Roman calendar, the points of the compass, and the same in Latin, in concentric circles. Rhumb lines are drawn from numerous centres over the surface of the terrestrial globe. The equator, ecliptic, and polar circles are painted boldly; while the parallels of latitude and meridians, at every ten degrees, are very faint lines.

The globe received additions, including the discoveries of Barents in Novaya Zemlya, and the date has been altered with a pen from 1592 to 1603. The constellations and fixed stars on the celestial globe are the same as those on the globe of Mercator, except that the Southern Cross has been added. On both the celestial and terrestrial globes of Molyneux there is a square label with this inscription:—

> "This globe belonging to the Middle Temple was repaired in the year 1818 by J. and W. Newton, Globe Makers, Chancery Lane."

1 The largest that had been made up to the time of their publication. The Behaim globe was 21 inches, the Mercator globes 16 inches, the Ulpius globe 15½ inches, and the Schoner globe 10½ inches in diameter. The others, which are older than the Molyneux globes, are very small. The diameter of the Laon

Over North America are the arms of France and England quarterly; supporters, a lion and dragon; motto of the garter; crown, crest, and baldrequin; standing on a label, with a long dedication to Queen Elizabeth.

The achievement of Mr. William Sanderson is painted on the imaginary southern continent to the south of Africa. The crest is a globe with the sun's rays behind. It stands on a squire's helmet with baldrequin. The shield is quarterly: 1st, *paly of six azure and argent, over all a bend sable* for Sanderson; 2nd, *gules, lions, and castles in the quarters* for Skirne *alias* Castilion; 3rd, *or, a chevron between* 3 *eagles displayed sable, in chief a label of three points sable* for Wall; 4th, *quarterly, or and azure, over all a bend gules* for Langston. Beneath there is an address from William Sanderson to the gentle reader, English and Latin, in parallel columns.

In the north polar regions there are several new additions, delineating the discoveries of English and Dutch explorers for the first time. John Davis wrote, in his *World's Hydrographical Discovery*: "How far I proceeded doth appear on the globe made by Master Emerie Molyneux." Davis Strait is shown with all the names on its shores which were given by its discoverer, and the following legend: "*Joannes Davis Anglus anno* 1585-86-87 *littora Americæ circum spectantia a quinquagesimo quinto grado ad* 73 *sub polarem scutando perlegit.*" On

globe is 6½ inches, of the Nancy globe 6 inches, and of the Lenox globe only 4½ inches.

another legend we have, "*Additions in the north parts to* 1603"; and below it are the discoveries of Barents, with his Novaya Zemlya winter quarters—"*Het behouden huis.*" Between Novaya Zemlya and Greenland there is an island called "*Sir Hugo Willoghbi his land*". This insertion arose from a great error in longitude, Willoughby having sighted the coast of Novaya Zemlya ; and the island, of course, had no existence, though it long remained on the maps. To the north of Siberia there are two legends—"*Rd. Cancelarius et Stephanus Burrow Angli Lappiæ et Coreliæ oras marinas et Simm. S. Nicolai vulgo dictum anno* 1553 *menso Augusto exploraverunt*"; and "*Joannes Mandevillanus eques Anglius ex Anglia anno* 1322 *Cathaiæ et Tartari regiones penetravit.*"

Many imaginary islands, in the Atlantic, are retained on the Globe: including "*Frisland*", "*Buss Ins*", "*Brasil*", "*Maidas*", "*Heptapolis*", "*St. Brandon*". On the eastern side of North America are the countries of Florida, Virginia, and Norumbega ; and also a large town of Norumbega up a gulf full of islands. The learned Dr. Dee had composed a treatise on the title of Queen Elizabeth to Norumbega ; and in modern times Professor Horsforth has written a memoir to identify Norumbega with a site up the Charles river, near Boston. On the Atlantic, near the American coast, is the following legend: "*Virginia primum lustrata, habitata, et culta ab Anglis inpensis D. Gualteri de Ralegh Equitis Aurati anmenti Elizabethæ In Angliæ*

Reginæ." On the western side of North America are California and Quiriua of the Spaniards, and Nova Albion discovered by Drake.

A legend in the Pacific Ocean furnishes direct evidence that information, for compiling the Globe, was furnished by Sir Walter Raleigh. It is in Spanish : "*Islas estas descubrio Pedro Sarmiento de Gamboa por la corona de Castilla y Leon desde el año* 1568 *llamolas Islas de Jesus aunque vulgarmente las llaman Islas de Salomon.*" Pedro de Sarmiento was the officer who was sent to fortify the Straits of Magellan after Drake had passed through. He was taken prisoner by an English ship on his way to Spain, and was the guest of Raleigh in London for several weeks, so that it must have been on information communicated by Raleigh that the statement respecting Sarmiento on this legend was based.

Besides "*Insulæ Salomonis*" there are two islands in the Pacific—"*Y Sequenda de los Tubarones*" and "*San Pedro*", as well as the north coast of New Guinea, with the names as given on Mercator's map.

Cavendish also appears to have given assistance, or possibly Molyneux himself accompanied that circumnavigator in his voyage of 1587. The words of a legend off the Patagonian coast seem to countenance this idea. They are : "*Thomas Caundish* 18 Dec. 1587 hæc terra *sub nostris oculis* primum obtulit sub latitud 47 cujus seu admodum salubris Incolæ maturi ex parte proceri sunt gigantes et vasti magnitudinis." The great southern continent is made to include Tierra del Fuego and the south coast of Magellan's Strait, and extends over the greater part of the south frigid zone.

S. Matheo, an island in the Atlantic, south of the line, was visited by the Spanish ships under Loaysa and Sebastian del Cano, but has never been seen since. It appears on the Globe. In the south Atlantic there are painted a sea-serpent, a whale, Orpheus riding on a dolphin, and ships under full sail—fore and main courses and topsails, a sprit sail, and the mizzen with a long lateen yard.

The tracks of the voyages of Sir Francis Drake and Master Thomas Cavendish round the world are shown, the one by a red and the other by a blue line. That these tracks were put on when the Globe was first made is proved by the reference to them in Blundeville's *Exercises.*

The name of the author of the Globe is thus given: "*Emerum Mullineux Angl. sumptibus Gulielm Sanderson Londinensis descripsit.*"

On the Celestial Globe there are the same arms of Sanderson, the same label by Newton, 1818, a briefer dedication to the Queen, date 1592, and "*Judocus Hondius Fon Sc.*" It would appear, therefore, that, when Molyneux had prepared the manuscript gores, they were entrusted to Hondius, the celebrated engraver and cartographer at Amsterdam, to print. A number of the globes were manufactured and sold; and some were made on a smaller scale, to serve for a cheaper edition.[1] Yet only one set has been preserved. It is in the library of the Middle

[1] See page 16.

Temple, and is the property of the Benchers of that Inn. This is certainly a strange depository for geographical documents of such interest and importance ; and it becomes a curious question how these globes, which would be so valuable to geographical and naval students, have found a final resting-place among the lawyers.

It is probable that they once belonged to Robert Ashley, who left his books to the Middle Temple, and whose portrait hangs in the library. This gentleman was descended from those of his name settled at Nashill, in Wiltshire. His father, Anthony Ashley,[1] married Dorothy Lyte, of Lytes Carey, in Somersetshire; and Robert was born at Damerham, seven miles from Salisbury, in 1565. He was at school at Southampton, under the well-known Master, Hadrian Saravia; and, as a boy, he had read *Bevis of Hampton*, *Guy of Warwick*, *Valentine and Orson*, *Arthur and the Knights of the Round Table*. When rather older, he perused the *Decameron*, and the *Heptameron* of the Queen of Navarre. In 1580 he went to Oxford, and in due time became a Barrister of the Middle Temple. Robert Ashley was an ardent geographer, and a very likely man to be the possessor of a set of the Molyneux Globes. He studied languages, and was

[1] Not to be confounded with Sir Anthony Ashley, who was at the sack of Cadiz, under the Earl of Essex, was Clerk to the Privy Council, and translated the *Mariner's Mirror* of Lucas Jansz Wagenaar into English in 1588. This Sir Anthony is the ancestor of the Earls of Shaftesbury.

master of French, Spanish, Italian, and Dutch. Fond of history and topography, he travelled over a great part of Europe, making the chambers in the Middle Temple his head-quarters. Ashley was an indefatigable collector, and made several translations.[1]

He lived amongst his books in the Temple almost entirely during the latter years of his long life. Ashley reached the age of seventy-six, dying in October 1641. He was buried in the Temple Church, and, by his will, the old Templar left all his books to the Inn in which he had dwelt so long. In April 1642 there was an order from the Benchers that the books left by Master Ashley should be kept under lock and key until a library was built. Thus Ashley's library formed the original nucleus of that of the Middle Temple. It contained a number of works on cosmography, including copies of two editions of the *Tractatus* on the Molyneux Globes by Hues. It is, therefore, highly probable that the globes themselves were included in Ashley's library, and that it was in this way that they found a last resting-place—one may almost say a burial-place—in the library of the Middle Temple.

[1] *Relation of the Kingdom of Cochin China* (1633, Bodleian, 4to.), from an Italian relation by Chr. Borri. *Uranie, or the Celestial Muse*, translated from the French of Bartas (1589). *Almansor, the Learned and Victorious King that Conquered Spain* (1627), from the edition printed at Salamanca in 1603. The Arabic original was in the Escurial, where Ashley saw it. A translation from the Italian of *Il Davide Persequitate* of Malvezzi (1637).

Almost as soon as the globes made their appearance, a manual for their use was published by Dr. Hood, of Trinity College, Cambridge, who gave lectures on navigation at Sir Thomas Smith's house in Philpot Lane.[1] In 1594 they were described by Blundeville in his *Exercises*, and in the same year a manual for their use was published in Latin by Robert Hues. The *Tractatus de Globis* of Hues passed through several editions, and as it has now been decided that it shall form one of the volumes of the Hakluyt Society, it will be well that a biographical notice of the author should precede the enumeration of former editions of his work.

Robert Hues (or Husius) was born in 1553, in a village called Little Hereford (pronounced Harford), in Herefordshire, eight miles north-east of Leominster. The parish is separated from Worcestershire by the river Teme. The church, dedicated to St. Mary Magdalene, is an ancient stone building in the Norman transition style, but unfortunately the registers only commence in 1697, and throw no light on the parentage of the great mathematician. He was well grounded at some local school, before he was sent up to Brasenose College at Oxford in 1571, where he was among the Servitors—"Pauperes Scholares". Here he continued for some time, as a very sober and serious student, but afterwards

[1] "*The Use of both the Globes, Celestial and Terrestrial, most plainly delivered in form of a dialogue. D. Hood, Mathematical Lecturer in the City of London, Fellow of Trinity College, Cambridge.*" (London, 1592, not paged. Bound up with Hues.)

removed to St. Mary Hall, taking his degree in about 1578. He was then noted for a good Greek scholar, and he is mentioned by Chapman as his learned and valued friend, to whose advice he was beholden in his translation of Homer.[1]

Hues appears to have travelled on the Continent soon after he took his degree, and on his return he devoted himself to the study of geography and mathematics, becoming well skilled in those sciences. He also made at least two voyages across the Atlantic, ·both probably with Thomas Cavendish. He mentions having observed for variation off the coast of North America[2]; so that he may have been with Cavendish when that navigator went with Sir Richard Grenville to Virginia. We learn from his epitaph that he accompanied Cavendish, and he himself says that he was sailing in the southern hemisphere in the years 1591 and 1592.[3] He must, therefore, have been on board the *Leicester* in the last voyage of Cavendish. It was a rough experience—gales of wind and wild weather in the Straits of Magellan, privations and hardships of all kinds, and on the passage home Cavendish died, and was buried at sea. Hues twice refers to the observations he made in this voyage in his *Treatise on the Globes*.[4] He must have returned to England just at the time when the Molyneux Globes were published, and

[1] Warton, *History of English Poetry*, iii, p. 442.

[2] See p. 121.

[3] See p. 66.

[4] Pp. 66, 67, 121.

his manual was written in the following year, and published in 1594.

The Oxford student had now added practical experience at sea to his theoretical knowledge. He had seen and observed the Southern Cross and the other stars of the Southern Hemisphere. He had ascertained the variation of the compass in the north, on the equator, and in the far south. He had acquired a knowledge of the requirements of navigators, and his *Tractatus de Globis* was intended to supply them with practically useful information. His *Breviarium Totius Orbis* was designed with the same object, and also went through several editions.

Henry Percy, Earl of Northumberland, granted a yearly pension to Robert Hues for the encouragement of his studies; and the accomplished scholar acted, for a year or two, as tutor to the Earl's son, Algernon, at Christ Church. During Northumberland's long and unjust imprisonment in the Tower he was solaced by the companionship of learned men, among whom were Thomas Heriot and Robert Hues; who also imparted their knowledge to Sir Walter Raleigh. Hues was one of Raleigh's executors. During the last years of his life Robert Hues resided almost entirely at Oxford, and there he died, in his eightieth year, on the 24th of May 1632, in the "Stone House", then belonging to John Smith, M.A., son of J. Smith, the cook of Christ Church. He was buried in Christ Church Cathedral, and a brass plate was put up to his memory, with the following inscription:—

"Depositum viri literatissimi, morum ac religionis integerrimi, Roberti Husia, ob eruditionem omnigenem, Theologicam tum Historicam, tum Scholasticam, Philologicam, Philosophiam, præsertim vero Mathematicam (cujus insigne monumentum in typis reliquit) Primum Thomæ Candishio conjunctissimi, cujus in consortio, explorabundis velis ambivit orbem: deinde Domino Baroni Gray; cui solator accessit in arca Londinensi. Quo defuncto, ad studia Henrici Comitis Northumbriensis ibidem vocatus est, cujus filio instruendo cum aliquot annorum operam in hac Ecclesia dedisset, et Academiæ confinium locum valetudinariæ senectuti commodum censuisset; in ædibus Johannis Smith, corpore exhaustus, sed animo vividus, expiravit die Maii 24, anno reparatæ salutis 1632, ætatis suæ 79."[1]

The first edition of the *Manual for the Globes*, by Robert Hues, is in the British Museum, and also at the Inner Temple. *Tractatus de Globis et eorum usu, accomodatus iis qui Londini editi sunt anno* 1593 (London, T. Dawson, 1594, 8vo.).

The second was a Dutch translation, printed at Antwerp. *Tractaut of te handebingen van het gebruych der hemel siker ende aertscher globe.*

[1] Wood's *History and Antiquities of Oxford* was written in English; bought by the University, in 1670, for £100, and published in Latin under the superintendence of Dr. Fell and the Curators of the Printing Office. Many things were altered, and there were some additions. *Historia et Antiquitatis UniversitatisOxoniensis duobus voluminibus comprehensæ* (Oxon., 1674), folio. Translation, 1786, 4to. The inscription is in the Latin edition (ii, p. 534). Under St. Mary Hall there is a notice of the death of Hues:—"Oxonii in parochiâ Sancti Aldati, inque Domicilio speciatim lapides, e regione insignis Afri cærulei, fatis concessit, et in ecclesiâ Ædis Christi Cathedrali humatus fuit an: dom: CIↃDXXXII (ii, p. 361) In laminâ œneâ, eidem pariati impactâ talem cernis inscriptionem" (ii, p. 288).

(Amb., 1597, 4to.). There are copies at the Universities of Louvain and Ghent.

The third is a reprint of the first edition, published at Amsterdam in 1611 (*Iudocus Hondius*, 8vo.). There are copies in the British Museum and Inner Temple.

The fourth reprint was in Dutch, also published at Amsterdam, *Tractaet of te handebingen van het gebruych der hemelsike ende aertscher globe* (Amstelodami, 1613, 4to.). A copy exists in the Royal Library at Brussels.

The fifth reprint appeared at Heidelberg in 1613, and contains the Index Geographicus. There are copies at the British Museum and in the Temple Library.

The sixth appeared at Amsterdam. *Tractatus de globis cœlesti et terrestri æorumque usu* (Amst., *Iudocus Hondius*, 1617, 4to.). There are copies at Louvain, Ghent, and Liege.

The seventh reprint was in a French translation by M. Haurion—*Traité des globes et de leur usage, traduit par Haurion* (Paris, 1618, 8vo.). There are copies in the Library of the Middle Temple, and at Louvain, Ghent, and Namur.

Of the eighth edition, published by Hondius at Amsterdam in 1624, in 4to., there are copies in the British Museum and at the Temple.

The ninth edition was published at Frankfort in 1627, in 12mo. There is a copy in the Musée Plantin at Antwerp.

The tenth edition is an English translation. *A Learned Treatise of Globes, both Cœllestial and Terres-*

triall, written first in Latin .... afterwards illustrated with notes by I. I. Pontanus, and now made English by J. Chilmead (London, 1638). Copies at the British Museum and in the Temple. The translator, John Chilmead, was of Christ Church College at Oxford. It is generally supposed that the name *John* was printed on the title-page in error, and that the translator was really Edmund Chilmead, who was born at Stow-in-the-Wold in Gloucestershire in 1610. This Edmund graduated in 1628, and was a Chaplain of Christ Church. Having been ejected in 1648 as a Royalist, he got his living in London by making translations and teaching music. He died in 1653, and was buried in the churchyard of St. Boltoph's Without, Aldersgate. Among his translations were the *Erotomania* of Ferrand, and a work on the Jews by Leo Modena; and he assisted in the translation of Procopius by Sir Henry Holbrooke. He also wrote a treatise on the music of the Greeks, which was printed at the end of the Oxford edition of Aratus, of 1672; and another on sound, which was never published.

The translation of the *Tractatus de Globis* of Hues certainly has *John* Chilmead on the title-page; but it is usually attributed to Edmund, and, as no *John* Chilmead, who was a translator and man of letters, is known to have lived at that time, the attribution is probably correct. But it is certainly a strange error to have made.

A Latin version of the *Tractatus de Globis* of Hues, by Jod. Hondius and I. I. Pontanus, appeared in London in 1659 (8vo.). There is a copy in the British Museum.

The twelfth edition of the work, and the second of the English version, with the notes of Pontanus, appeared in London in 1659 (8vo.). There is a copy in the Library of Sion College.

The last edition of the Latin version was published at Oxford in 1663. There is a copy in the Bodleian Library.

I. Isaac Pontanus, who annotated the Amsterdam editions of the *Tractatus de Globis*, and whose notes were translated for the English editions, was a cosmographer and historian of great eminence. He was the son of a merchant originally from Haarlem, who was Consul at Elsinore for the States-General. Pontanus was born while his parents were residing at Elsinore, on the 21st of January 1571. For three years he was the pupil of Tycho Brahe, on the Island of Hveen, and he always retained a feeling of profound veneration for his illustrious master. He afterwards studied at Basle and Montpellier. On his return to Holland he was appointed Professor of Philosophy and History in the College of Harderwyck, a post which he retained until his death, and in 1620 he was nominated Historiographer to the King of Denmark. He wrote many learned works, including a ponderous Danish history[1]; but his most valuable contribution to geographical literature was his *History of Amsterdam*.[2] Pontanus was a constant

[1] *Rerum Danicarum Historia* (Amst., 1631).

[2] *Historia urbis et rerum Amstelodamensium* (Amst., 1611).

advocate of exploring enterprise, and gave much assistance to the cartographer Hondius in his arduous undertakings. Owing to his profound learning, the deep interest he took in the science of navigation, and his knowledge of mathematics, no better editor of the Dutch editions of the work of Hues could have been found than Isaac Pontanus. He died at Harderwyck on the 6th of October 1639, aged 68.

Hues opened his work with an epistle dedicatory to his intimate friend, Sir Walter Raleigh[1]; in which he recapitulated the discoveries made by Englishmen during the reign of the great Queen ; and urged that his countrymen would already have surpassed the Spaniards and Portuguese, if they had taken more pains to acquire a complete knowledge of geometry and astronomy. The efforts of Englishmen, he believed, had been rendered less effective, owing to their ignorance of the sciences, a knowledge of which is essential to a successful navigator. He concluded by saying that he had composed his treatise in the hope that it might be useful in advancing a study of the seaman's art. In his Preface, Master Hues went to the root of the matter, and proceeded to prove the sphericity of the earth ; first advancing the usual arguments, and then refuting the theories of those who disputed them. He devoted some space to those who argued that the mountains prevented the earth's surface from being

[1] The opening lines of the address, and the name of Sir Walter Raleigh, are omitted in the English translations.

round; and to others who maintained that a liquid surface is flat and not concave. Having established his points, the conclusion that a globe is the best form by which to represent a spherical body was inevitable. He concluded with some remarks in commendation of the Molyneux Globes, constructed through the liberality of Master Sanderson. They are more than twice the size ot Mercator's globes, which is a great advantage; and they contained all the most recent discoveries.

The treatise itself is divided into five parts, the first treating of things which are common to both globes; the second devoted to the planets, fixed stars, and their constellations; the third to a description of land and sea portrayed on the Terrestrial Globe, and to a discussion respecting the circumference of the earth; and the fourth explains the use of the globes. The fifth part consists of a learned treatise by Master Herriot on the rhumb lines and their uses.

In the first part the frame is described, on which the globe is set; the broad wooden horizon, with its various divisions; and the brass meridian at right angles to it, on the poles of which the globe itself is fixed. The *Horarius* is a small circle of brass, divided into twenty-four equal parts, to be fixed on one of the poles of the meridian with a pin, called the *Index Horarius*, made to point to each of the twenty-four divisions as the globe turns on its axis. Having described these accessories of the globe, Hues next turns to the circles and lines drawn on the globe itself, discussing questions relating to them in very full detail, and also treating of the zones and climates. His frequent references to the theories and calculations both of the ancients and of his contemporaries give that kind of biographical interest to his dissertations which serves, better than any other method, to impress scientific facts on the memory.

The second part treats of the celestial globe and of the Ptolemaic constellations and stars, with the stories of the origin of their Greek names, and of those adopted, in later days, by the Arabian astronomers. Pontanus, in his foot-notes, brings our thoughts back to the supposed double origin of the constellations in the remotest antiquity.[1] He suggests that the ideas were conceived, and the names given, by two classes of men, the sailors of the Phœnician coasts and the husbandmen of the Chaldean plains. It was a more modern theory that some of the constellations, derived from the Phœnicians, represented the figure-heads of ships, or the emblematic replicas of them hung up in the temples; such as Aries, Taurus, Pegasus, Cygnus, Hydra, Cetus, Delphinus. Taurus and Pegasus are actually represented as half figures, just as figure-heads would be. The most ancient constellations, the *Geniculator*, or man doomed to labour on his knee (converted by the Greeks into *Hercules*), the Nimrod or Orion, the Centaur, and the Serpentarius were, it is supposed, of Chaldean origin.

[1] Notes, pp. 49 and 59.

Sometimes both the names given by the sailors and those of the shepherds were continued, as in the case of the Bear, also known as the Waggon or Chariot. Pontanus, in his foot-notes, twice refers to the passages in the book of Job where certain Hebrew words are translated as stars—Arcturus, Orion, the Pleiades, and Mazzaroth; but the idea that the equivalent Hebrew words have any allusion to stars is a mere conjecture, and, it would seem, an improbable one.[1]

The immense antiquity of the names for constellations is proved by the lines in Homer:

> "The Pleiads, Hyads, with the northern team,
> And great Orion's more effulgent beam,
> To which, around the axle of the sky,
> The Bear revolving, points his golden eye,
> Still shines exalted in the ethereal plain,
> Nor bathes his blazing forehead in the main."
> (Pope's *Iliad.*)

---

1 "Which maketh Arcturus (*Ash*), Orion (*Kesil*), and Pleiades (*Kimah*), and the chambers of the south." (Job ix, 9.)

"Canst thou bind the sweet influences of Pleiades (*Kimah*) or loose the bands of Orion (*Kesil*). Canst thou bring forth *Mazzaroth* in his season or canst thou guide Arcturus (*Ash*) with his sons?" (Job xxxviii, 31, 32.)

"Seek him that maketh the seven stars and Orion, and turneth the shadow of death into the morning." (Amos v, 8.)

In a foot-note (p. 52), Pontanus discusses the name of Arcturus, and mentions that the word which is given as Arcturus in the Septuagint is *Ash* in Hebrew, from the root *Grusch*—"congregabit". *Ash* is also translated as "vapour", *Kesil* as "cold" or "snow" ("rage" or "madness", according to Pontanus), and *Kimah* as "rain". *Mazzaroth*, a periodical pestilential wind. No similar words are used for stars by the Arabian astronomers; and it is supposed, by some authorities, that no reference to stars was intended either in Job or Amos.

This passage shows that the constellations in the days of Homer were the same as those enumerated in the poem of Aratus, who is constantly referred to by Hues. Ptolemy adopted the names in Aratus, and thus they have been transmitted, through the Arabs, to modern times. In this second part our author passes them all in review, with their Arabic names, here and there noticing the assertions and theories of later or contemporaneous writers, such as Cardan, Patricius, and Corsalius. In correcting the errors of some of these authors, based on the vague narrative of Amerigo Vespucci, Hues takes occasion to give his impressions of the stars in the southern hemisphere, derived from a severe service of more than a year in those seas, on board the *Leicester*, with Cavendish.[1] The second part of the *Tractatus* supplied an admirable explanatory guide to the Celestial Globe.

In the third part Hues undertook to describe the lands and seas delineated on the Terrestrial Globe. He begins by explaining the ideas respecting the three continents of the old world which were entertained by the ancients, and shows how these early speculations were corrected by subsequent discoveries. He then reviews the bounds of the knowledge of his own times, when the northern limits had been extended to 73°, with fair hopes that the ocean bounds the northern shores of America; and the south had been made known as far as the Straits of Magellan. He evidently inclined to a belief in a vast

[1] Pp. 66, 67.

southern continent, such as is delineated on the globe. Next, he enumerates the countries contained in the four continents; and refers to the unknown regions of Australia to the south of New Guinea, and to the vast tracts in the far north, which then, as now, remain to be discovered. But this part of his work is confessedly incomplete, and in his preface he refers his readers to the more detailed information given by Ortelius and Mercator.

In a second chapter of his third part Hues discusses the various methods that had been adopted to ascertain the circumference of the earth and the length of a degree. He gives an interesting account of the labours of Eratosthenes and Posidonius; and as the great differences in the results of various ancient authorities were partly due to the standards of measurement, he devotes some space to a discussion of the various lengths given to a degree.

The fourth part of the *Tractatus*, in which the practical uses to which the Globe may be put by the navigator are described, was the most important in the eyes of the author, and the one by means of which he hoped to be of most service to his countrymen. Previous to the discovery of logarithms, the problems of nautical astronomy could only be worked out with the help of very prolix mathematical calculations by practical scholars. But the globe supplied methods of finding the place of the sun, latitude, course, and distance, amplitudes and azimuths, time and declination, by inspection. This was a great boon to navigation, and the globe came into very general use on board ship. As a practical guide to its use the treatise of Hues became a most valuable book to sailors; so that it played no unimportant part in furthering the exploring enterprises of Englishmen in the seventeenth century.

The fourth part opens with a definition of longitude, and the various ways of finding it. Observations of eclipses of the moon are pronounced to be the most accurate method, but one very seldom used. As to proposals for finding longitude by observations of differences of time, with clocks or hour-glasses, Hues scouts the idea, which had been rejected by all learned men; the clocks of that period being altogether unable to perform that which was required over them. Navigators would have to wait for nearly two centuries before mechanical skill had reached to the height of constructing a chronometer. Meanwhile, the substitutes were worthless, and those who sold them were impostors. "Away," cried Mr. Hues, "with all such trifling, cheating rascals!" As regards latitude Hues reminds his readers that it is always the same as the height of the pole above the horizon, a measurement which was easily made. He then explains the methods of using the globe for finding the altitude of a heavenly body, its place and declination, the latitude by meridian altitude, the right ascension of heavenly bodies, their azimuths and amplitudes, the time and duration of twilight, the variation of the compass, and how to make a sun dial by the globe.

The fifth part is a valuable treatise by Thomas Herriot,[1] another eminent mathematician, on the

[1] Thomas Herriot was born at Oxford in 1560, was a Commoner of St. Mary Hall, and took his M.A. degree in 1579. He was an excellent mathematician, and was employed by Sir Walter Raleigh to instruct him in that science, becoming a member of his family for some time. When Raleigh fitted out the expedition to Virginia, under Sir Richard Greville, in 1585, young Herriot became a member of it, and made a map of the country. On his return he published a *Brief and True Report of the new found land of Virginia* which was reprinted by Hakluyt. Herriot devoted himself to mathematical studies, especially to algebra, and was also an astronomer and a practical navigator. Raleigh introduced him to the Earl of Northumberland, who gave him a pension of £120 a year, and he resided for some time at Sion College. When Northumberland was committed to the Tower, Thomas Herriot, with his learned friends, Robert Hues and Walter Warner, solaced his long imprisonment by their conversation. They were called the Earl's three Magi. Herriot corresponded with Kepler on the theory of the rainbow. He died on July 2nd, 1621, of a cancer on the lip; and was buried in St. Christopher's Church, where there was a monument to his memory, with the following inscription:

" Siste viator, leviter preme,
Jacet hic juxta quod mortale fuit
C. V.
Thomæ Harrioti
Hic fuit doctissimus ille Harriotus
de Syon ad flumen Thamesin
Patria et educatione
Oxoniensis
Qui omnes scientias calluit
Qui in omnibus excelluit
Mathematicis, Philosophicis, Theologicis
Veritatis indagator studiosissimus
Dei Trini unius cultor piissimus
Sexagenarius aut eo circiter
Mortalitati valedixit, Non vitæ
Anno Christi MDCXXI, Julii 2."

rhumb lines described on the Terrestrial Globe, and their uses. Herriot shows that five nautical problems may be solved by the rhumb lines, and that if any two of the four elements—course, distance, diff. long., and diff. lat.—are known, the other two can be found. Each of these five problems is given, with a practical example; and the only one which presented serious difficulty is that in which it is required to find the course and diff. lat. when diff. long. and distance are given. This cannot be puzzled out on the globe without long and tedious calculation, and even then the result is useless.

The *Index Geographicus* is only given in one or two editions. It is a long and very complete list of places, with their latitudes and longitudes as shown on the globe. The list may often be useful to geographical students, as a help towards the identification of old names, or of names made obscure by peculiar spellings, and it has, therefore, been thought desirable that it should be reprinted.

The only foot-notes to the text are those referring to the annotations of Pontanus in the Amsterdam editions. Information respecting the names of astronomers and others mentioned in the text, the stars and constellations, the names of places, and scientific terms will be found in the Indices. The Biographical Index contains short notices of astronomers and mathematicians, as well as references to the places in the text where their names occur. The Astronomical Index, for most valuable help in the preparation of which I am indebted to Professor Robertson

Smith of Cambridge, has been prepared on the same plan. The Index of Names of Places, and that of Scientific Terms, are merely intended for furnishing references to the pages in the text.

(Latin Title.)

TRACTATVS DE GLOBIS

ET EORVM VSV,

Accomodatvs iis qui Londini editi svnt anno 1593,

Sumptibus Guglielmi Sandersoni
Ciuis Londinensis,

Conscriptvs à

ROBERTO HUES.

Londini
In ædibvs Thomæ Dawson.
1594.

(English Title.)

A LEARNED TREATISE OF Globes,

Both Cœlestiall and Terrestriall: with their ſeveral uſes.

Written first in Latine, by Mr Robert Hues: and by him ſo Published.

Afterward Illustrated with Notes, by Io. Iſa. PONTANUS.

And now lastly made English, for the benefit of the Vnlearned.

By John Chilmead Mr A. of Christ-Church in Oxon.

LONDON,
Printed by the Aſſigne of T. P. for P. Stephens and C. Meredith, and are to be sold at their Shop at the Gold[en Li]on in Pauls-Church-yard. 1[638.]

N.B.—Letters within brackets torn out of original; the date, also torn out, is given at the end of the work.

THE CONTENTS OF THE CHAPTERS OF THIS TREATISE.

CHAPTER IV.

Of the Zones and their number.—The vaine opinions of the Ancients, concerning the temperature of the Zones, are rejected ; both by the Testimonies of some of the Ancients themselves, as also by the experience of later times.

CHAPTER V.

Of the Amphiscij, Periscij, and Heteroscij.

CHAPTER VI.

Of the Periæci, Antæci, and Antipodes compared to each other.

CHAPTER VII.

Of Climates and Parallels.

THE SECOND PART.

CHAPTER I.

Of such things as are proper to the Cœlestiall Globe ; as namely of the Stars. And first of the Planets, or Wandering Stars.

CHAPTER II.

Of the fixed Starres and their Constellations.

CHAPTER III.

Of the Constellations of the Northerne Hemisphære.

CHAPTER IV.

The signes of the Zodiacke ; and first of the Northerne.

CHAPTER V.

The Constellations of the Southerne Hemisphære ; and first of those in the Zodiacke.

CHAPTER VI.

Of the rest of the Constellations of the Southerne Hemisphære.

CHAPTER VII.

Of the other Stars which are not expressed in Globes.—Why the Stars appeare sometimes in greater numbers than at other times, and sometimes greater and at other times lesse ; with the confutation of some vain opinions concerning the same.—The idle relations of Americus Vespasius, Cardan, and Patricius concerning the extraordinary greatnesse of the Stars about the South Pole are refuted out of the Author's own experience.

THE THIRD PART.

CHAPTER I.

The Geographicall description of the Terrestriall Globe, with the parts of the world that are yet knowne. The errours of Ptolomy concerning the Southerne bounds of Africa and Asia, as also of the Northerne limits of Europe, are condemned out of the writings of the Ancients and various experience of later Writers.

CHAPTER II.

Of the compasse of the Earth and the measure of a degree : with diverse opinions concerning the fame of the Greeks ; as namely, Eratosthenes, Hipparchus, Posidonius, Cleomedes, and Ptolomy ; as also of the Arabians, Italians, Germans, English, and Spanish.—Posidonius and Eratosthenes are confuted out of their owne observations and propositions. Ptolomyes opinion is preferred before the rest, and he freed from the calumnies of Maurolycus ; who is also taxed in that without cause favouring Posidonius hè unjustly condemns Ptolomy.

THE FOURTH PART.

CHAPTER I.

How to finde out the longitude, latitude, distance and angle of position or situation of any places expressed in the Terrestriall Globe.

CHAPTER II.

Of the Latitude of any place.

CHAPTER III.

How to finde the distance and angle of position of any two places.

CHAPTER IV.

To finde the Altitude of the Sunne or Starres.

CHAPTER V.

To finde the place and declination of the Sunne for any day given.

CHAPTER VI.

To finde the Latitude of any place by observing the Meridian altitude of the Sunne or Starres.

CHAPTER VII.

How to finde the Right and Oblique Ascension of the Sunne and Starres for any latitude of Place and Time.

CHAPTER VIII.

How to finde the Horizontall difference betwixt the Meridian and the verticall circle of the Sunne or any other Starre which they call the Azimuth, for any time or place assigned.

MEMORANDUM.

Postremo est tabula Geographica in qua Regionum, Insularum, fluviorii, Promontoriorum, Sinnum, Montium & reliquarum quæ in Terrestri Globo exprimuntur, nomina omnia ordine Alphabetico digesta est: adjecta singulis sua longitudine & latitudine.

To the most illustrious and honourable Sir Walter Raleigh, Knight, Captain of the Queen's Guard, Lord Warden of the Stannaries in the Counties of Cornwall and Devon, Vice-Admiral of Devon, Robert Hues wishes lasting happiness.

MOST ILLUSTRIOUS SIR,

That nothing is at once brought forth, and perfected, is an observation wee may make as from other things, so in a more especial manner from Arts and Sciences. For (not to speake anything of the rest which yet have all of them in succession of times had their accessions of perfection), if wee but take the astronomicall writings of Aratus, or of Eudoxus (according to whose observations Aratus is reported by Leontius Mechanicus to have composed his Phænomena), and compare the same with the later writings of Ptolomy: what errours and imperfections shall we meet withall?

And in the Geographicall workes of the Ancients, whether we compare them among themselves, the later with the former; or either of them with the more accurate descriptions of our Moderne Geographers: how many things shall we meet withall therein, that need either to be corrected as erroneous, or else supplied as defective? There shall wee finde Strabo everywhere harshly censuring the extravagances of Eratosthenes, Hipparchus, Polybius, and Posidonius: Authors among the Ancients of very high esteem. For as for Pytheas, Euthemeres, Antiphanes, and those Indian Historiographers Megasthenes, Nearchus, and Daimachus, whose writings are stuffed with so many fabulous idle relations, he accounts them unworthy of his censure. In like manner Marinus Tyrius, however a most diligent writer, is yet hardly dealt withall by Ptolomy. And even Ptolomy himselfe, a man that for his great knowledge and experience may seem to have excelled all those that went before him; yet, if a man shall but compare his Geographicall Tables with the more perfect discoveries of our later times, what defects and imperfections shall hee there discover?

Who sees not his errours in the bounds he sets to the Southern parts of Asia and Africa? How imperfect are his descriptions of the Northern coasts of Europe? These errours of Ptolomy and of the Ancient Geographers have now at length been discovered by the late Sea voyages of the Portugalls and English; the Southern coasts of Africa and Asia having beene most diligently searched into by the Portugalls as the Northerne parts of Europe have in like manner beene by our owne Country-men. Among whom the first that adventured on the discovery of these parts were, Sir Hugh Willoughby and Richard Chanceler, after them Stephen Borough. And further yet then either of these, did Arthur Pet and Charles Iackman discover these parts. And these voyages were all taken by the instigation of Sebastian Cabot; that so, if it were possible, there might be found out a nearer passage to Cathay and China: yet all in vaine; save only that by this meanes a course of trafficke was confirmed betwixt us and the Moscovite.

When their attempts succeeded not this way, their next designe was then to try what might be done on the Northerne coasts of America; and the first undertaker of these voyages was Mr. Martin Frobisher: who was afterwards seconded by Mr. Iohn Davis. By means of all which Navigation many errours of the Ancients, and their great ignorance, was discovered.

But now that all these their endeavours succeeded not, our Kingdome at that time being well furnished in ships and impatient of idlenesse, they resolved at length to adventure upon other parts. And first Sir Humphrey Gilbert with great courage and Forces attempted to make a discovery of those parts of America which were yet unknowne to the Spaniard, but the successe was not answerable. Which attempt of his was afterward more prosperously prosecuted by Sir Walter Rawleigh; by whose meanes Virginia was first discovered unto us; the Generall of his forces being Sir Richard Greenvile; which Countrey was afterwards very exactly surveighed and described by Mr. Thomas Hariot.

Neither have our country-men within these limits bounded their Navigations. For Sir Francis Drake, passing through the Straites of Magellane, and bearing up along the Westerne Coasts of America, discovered as farre as 50 degrees of Northerne Latitude. After whom Mr. Thomas Candish, tracing the same steps, hath purchased himselfe as large a monument of his fame with all succeeding ages. I shall not need to reckon with these our Countryman, Sir Iohn Mandevil, who almost 300 years since in a 33 years voyage by land took a strict view of all India, China, Tartary, and Persia, with the Regions adjoyning.

By these and the like expeditions by Sea, the matter is brought to that passe that our English Nation may seeme to contend even with the Spaniard and Portugall himselfe for the glory of navigation. And without all doubt, had they but taken along with them a very reasonable competency of skill in Geometry and Astronomy, they had by this gotten themselves a farre more honourable name at Sea than they. And, indeed, it is the opinion of many understanding men that their endeavours have taken the lesse effect meerely through ignorance in these Sciences. That, therefore, there might be some small accrument to their study and paines that take delight in these Arts, I have composed this small treatise, which that it may be for their profit I earnestly desire.

Farewell.

THE PREFACE.

THERE are two kinds of Instruments by which Artificers have conceived that the figure of this so beautifull and various fabricke of the whole Universe might most aptly be expressed, and as it were at once presented to the view. The one exhibiting this Idea in a round solid is called a Globe, or Sphære. The other, expressing the same in a Plaine, they tearme a Planisphære, or Map. Both of which having been long since invented by the Ancients have yet even to our times in a continued succession received still more ripenesse and perfection. The Sphære or Globe, and the use thereof, is reported by Diodorus Siculus to have been first found out by Atlas of Libya: whence afterward sprung the Fable of his bearing up the Heavens with his Shoulders. Others attribute the invention of the same to Thales. And it was afterward brought to perfection by Crates (of whom Strabo makes mention), Archimedes, and Proclus; but most of all by Ptolomy; according to whose rules, and observations especially, succeeding times composed their Globes, as Leontius Mechanicus affirmes. And now there hath been much perfection added to the same in these our later times by the industry and diligence of Gemma Frisius and Gerardus Mercator; as it may appear by those Globes that were set forth at London, Anno 1593, so that now there seemes not to be anything that may be added to them. The Planisphære, indeed, is a fine invention, and hath in it wonderfull varietie of workmanship, if so be that the composition of it be rightly deduced out of Geometricall and Opticall principles; and it wants not its great delightfulness and beauty also. But yet that Other, being the more ancient, hath also the priority in Nature, and is of the most convenient forme; and therefore more aptly accomodated for the understanding and fancy (not to speake any of the beauty and gracefulnesse of it), for it representeth the things themselves in proper genuine figures.

For as concerning the figure of the Heavens whether it was round was scarcely ever questioned by any. So likewise touching the figure of the earth, notwithstanding many and sundry opinions have been broached among the ancient Philosophers, some of them contending for a plaine, others an hollow, others a cubicall, and some a pyramidall forme; yet the opinion of its roundnesse with greatest consent of reason at length prevailed, the rest being all exploded. Now wee assume it to be round, yet so as that wee also admit of its inequalities, by reason of those so great eminences of hilles and depression of vallies. Eratosthenes, as he is cited
Strabo. by Strabo in his first books, saith that the fashion of the Earth is like that of a Globe, not so exactly round as an artificiall Globe is, but that it hath certain inequalities. The earth cannot be said to be of an exact orbicular forme,
Pliny. by reason of so many hilles and low plaines, as Pliny rightly observes. And Strabo, also, in his first book of his Geography, saith that the earth and the water together make up one sphæricall body, not of so exact a forme as that of the Heavens, although not much unlike it. This assertion of the roundnesse of the Earth with the intervening Sea is confirmed also by these reasons. For, first, that it is round from East to West is proved by the Sun, Moon, and the other Starres, which are seen to rise and set first with those that inhabite more Eastwardly, and afterward with them that are farther West. The Sun riseth with the Persians that dwell in the Easterne parts foure hours sooner than it doth with
Cleom., lib. 1. those that dwell in Spaine, more Westward, as Cleomedes affirms. The same is also proved by the observing of Eclipses,

especially those of the Moon, which, although they happen at the same time, are not yet observed in all places at the same houre of the day or night, but the houre of their appearing is later with them that inhabite Eastward then it is with the more Westerne people. An Eclipse of the Moon, which Ptolomy reports, lib. 1, *Geogr.*, Cap. 4, to have been in Arbela (a towne in Assyria) at the fift hour of the night, the same was observed at Carthage at the second houre. In like manner an Eclipse of the Sun, which was observed in Campania to be betwixt 7 and 8 of the clock, was seen by Corbulo, a Captain in Armenia, betwixt 10 and 11, as it is related by Pliny. Now that it is also of a sphericall figure from North to South may be clearly demonstrated by the risings, settings, elevations, and depressions of the Starres and Poles. The bright Starre that shines so resplendently in the upper part of the sterne of the Ship *Argo*, and is called by the Greeks *κὰνωβ*, is scarcely to be seen at all in Rhodes, unlesse it be from some eminent high place; yet the same is seen very plainly in Alexandria, as being elevated above the Horizon about the fourth part of a signe, as Proclus affirmes in the end of his book, *de Sphæra*. For I read it *Conspicue cernitur*, not as it is commonly, *Prorsus non Cernitur ;* notwithstanding that both the Greek text and also the Latine translation are against it. Another argument may be taken from the figure of the shadow in the Eclipse of the Moon, caused by the interposition of the Earth's opacous body; which shadow being sphæricall, cannot proceed from any other than a round globular body, as it is demonstrated unto us out of opticall principles. But this one reason is beyond all exception, that those that make toward the land at Sea shall first decry the tops of the hilles onely, and afterward, as they draw nearer to shore, they see the lower parts of the same by little and little, which cannot proceed from any other cause than the gibbositie of the Earth's superficies.

Ptolomy.

Proclus.

As for those other opinions of the hollow, cubicall,

pyramidall, and plaine figure of the Earth, you have them all largely examined both in Theon (Ptolomies Interpreter), Cleomedes, and almost in all our ordinary authors of the Sphære, together with the reasons why they are rejected. Yet that old conceit of the plainnesse of the Earth's superficies is againe now at last, *tanquam Crambe recocta*, set forth in a new dresse, and thrust upon us by Franciscus Patricius, who, by some few cold arguments and misunderstood experiments, endeavours to confirme his owne, and, consequently, to overthrow that other received opinion of the sphæricall figure of the Earth. I shall onely lightly touch at his chiefest arguments; my present purpose and intention suffering me not to insist long on the Confutation of them. And first of all the great height of hilles, and the depression of vallies, so much disagreeing from the evennesse of the plain parts of the Earth, seem to make very much against the roundnesse of the Earth. Who can heare with patience, saith he, that those huge high mountaines of Norway, or the mountain Slotus which lies under the Pole, and is the highest in the world, should yet be thought to have the same superficies with the Sea lying beneath it? This, therefore, being the chiefest reason that may seem to overthrow the opinion of the Earth and Seas making up one sphæricall body, let us examine it a little more nearly, and consider how great this inequality may be, that seemes to make so much against the evennesse of this Terrestriall Globe. Many strange and almost incredible things are reported by Aristotle, Mela, Pliny, and Solinus, of the unusuall height of Athos, an hill in Macedonia, and of Casius in Syria, as also of another of the same name in Arabia, and of the mountaine Caucasus. And among the rest one of the most miraculous things which they have discovered of the mountaine Athos is, that whereas it is situate in Macedony, it casts a shadow into the market-place at Myrrhina, a towne in the Island Lemnos, from whence Athos is distant 86

Mela, l. 1, c. 2. Sol., c. 17. Plin., l. 3, c. 12.

miles. But for as much as Athos lies westward from Lemnos, as may appeare out of Ptolomies Tables, no marvaile that it casts so large a shadow, seeing that wee may observe by daily experience, that as well when the Sun riseth as when it sets, the shadowes are always extraordinary long.

But that which Pliny and Solinus report of the same Mountaine I should rather account among the rest of their fabulous Stories, where as they affirme it to be so high that it is thought to be above that region of the Aire whence the rain is wont to fall. And this opinion (say they) was first grounded upon a report that there goes, that the ashes which are left upon the Altars on the top of this hill are never washed away, but are found remaining in heapes upon the same. To this may be added another testimony out of the Excerpts of the seventh book of Strabo, where it is said that those that inhabite the top of this Mountaine doe see the Sun three houres sooner than those that live near the Sea side. The height of the Mountaine Caucasus is in like manner celebrated by Aristotle, the top whereof is enlightened by the Sunnes beames the third part of the night, both morning and evening. No lesse fabulous is that which is reported by Pliny and Solinus of Casius, in Syria, from whose top the Sun rising is discovered about the fourth watch of the night; which is also related by Mela of that other Casius in Arabia. But that all these relations are no other than mere fables is acutely and solidly proved by Petrus Nonius out of the very principles of Geometry. As for that which Eustathius writes, that Hercules Pillars, called by the Greeks Calpe and Abenna, are celebrated by Dionysius Periegetes for their miraculous height, is plainly absurd and ridiculous. For these arise not above an hundred elles in height, which is but a furlong; whereas the Pyramids of Egypt are reported by Strabo to equall that height; and some trees in India are found to exceed it, if wee may credit the relations of those Writers who, in the same Strabo, affirme that there grows a

L. de Crepusculis.

Eustathius.

Strabo, lib. 15.

tree by the river Hyarotis that casteth a shadow at noon five furlongs long.

Those fabulous narrations of the Ancients are seconded by as vaine reports of our moderne times. And first of all Scaliger writes from other men's relations that Tenariff, one of the Canary Islands, riseth in height fifteene leagues, which amount to above sixtie miles. But Patricius, not content with this measure, stretcheth it to seventie miles. There are other hilles in like manner cryed up for their great height, as, namely, the Mountaine Andi, in Peru, and another in the Isle Pico, among the Azores Islands; but yet both these fall short of Tenariff. What credit these relations may deserve we will now examine. And first for Tenariff, it is reported by many writers to be of so great height that it is probable the whole world affordes not a more eminent place; not excepting the Mountaine Slotus itself, which, whether ever any other mortall man hath seen, beside that Monke of Oxford (who, by his skill in Magicke, conveighed himselfe into the utmost Northerne regions and tooke a view of all the places about the Pole, as the Story hath it), is more than I am able to determine. Yet that this Isle cannot be so high as Scaliger would have it we may be the more bold to believe, because that the tops of it are scarcely ever free from snow, so that you shall have them covered all over with snow all the year long, save onely one, or, at the most, two months in the midst of summer, as may appeare out of the Spanish Writers. Now that any snow is generated 60 or 70 miles above the plaine superficies of the Earth and Water is more then they will ever persuade us, seeing that the highest vapours never rise above 48 miles above the Earth, according to Eratosthenes his measure; but according to Ptolomy they ascend not above 41 miles. Notwithstanding, Cardan and some other profest Mathematicians are bold to raise them up to 288 miles; but with no small staine of their name have they mixed those trifles with their other writings. Solinus

Exer.. 43 Contra Card.

Card. de Subt. Theod. Win., par. 2, Spher. quaestionem.

reports that the tops of the Mountaine Atlas reacheth very neare as high as the circle of the Moon; but he betrayeth his own errour in that he confesseth that the top of it is covered with snow, and shineth with fires in the Night. Not unlike to this are those things which are reported of the same mountaine and its height by Herodotus, Dionysius Afer, and his scholiast Eustathius; whence it is called in Authours, Cœlorum Columen, the pillar that bears up the Heavens. But to let passe these vaine relations, let us come to those things that seem to carry a greater show of truth. Eratosthenes found by Dioptricall instruments, and measuring the distances betwixt the places of his observation, that a perpendicular drawn from the top of the highest mountaine down to the lowest bottome or vally, did not exceed ten furlongs. Cleomedes saith that there is no hill found to be above fifteene furlongs in height, and so high as this was that vast steepe rocke in Bactriana, which is called Sisimitræ Petra, mentioned by Strabo in his II booke of his Geography. The toppes of the Thessalian Mountaines are raised to a greater height by Solinus then ever it is possible for any hill to reach. Yet, if we may believe Pliny, Dicæarchus being employed by the king's command in the same businesse, found that the height of Pelion, which is the highest of all, exceeded not 1,250 pases, which is but ten furlongs. But to proceed yet a little further, lest we should seem too sparing herein, and to restraine them within narrower limits than wee ought, wee will adde to the height of hilles the depth also of the Sea. Of which the illustrious Iulius Scaliger, in his 38 exercitations against Cardan, writeth thus: The depth of the Sea (saith he) is not very great, for it seldome exceeds 80 pases, in most places it is not 20 pases, and in many places not above 6; in few places it reacheth 100 pases, and very seldome, or never, exceeds this number. But because this falles very short of the truth, as is testified by the daily experience of those that passe the Sea, let us

Theon. I, com. in. Ptol.

L. 1, c. 63.

make the depth of the Sea equall to the height of Mountaines: so that suppose the depth thereof to be 10 furlongs, which is the measure of the Sardinian Sea in the deepest places, as Posidonius in Strabo affirmes. Or, if you please,
Cleom. let it be 15 furlongs, as Cleomedes and Fabianus, cited by
Plinius. Pliny, lib. 2, c. 102, will have it. (For Georg. Valla, in his interpretation of Cleomedes, deales not fairely with his Authour, where he makes him assigne 30 furlongs to be the measure of the Sea's depth.) These grounds being thus laid, let us now see what proportion the height of hilles may bear to the Diameter of the whole Earth; that so we may hence gather that the extuberancy of hilles are able to detract little or nothing from the roundnesse of the Earth, but that this excrescency will be but like a little knob or dust upon a ball, as Cleomedes saith. For if wee suppose the circumference of the whole Earth to be 180,000 furlongs, according to Ptolomies account (neither did ever any of the Ancients assigne a lesse measure than this, as Strabo witnesseth), the Diameter therefore will be (according to the proportion betwixt a circle and its diameter found out by Archimedes) above 57,272 furlongs. If, then, we grant the highest hilles to be ten furlongs high, according to Eratosthenes and Dicæarchus, they will beare the same proportion to the Diameter of the Earth that is betwixt 1 and 5,727. (Peucerus mistakes himselfe when he saith that the Diameter of the Earth to the perpendicular of ten furlongs is as 18,000 to 1, for this is the proportion it beareth to the whole circumference, and not the diameter. Or suppose the toppes of the highest hilles to ascend to the perpendicular of fifteene furlongs, as Cleomedes would have it, the proportion then will be of one to 3,818. Or if you please let it be thirtie furlongs, of which height is a certain rock in Sogdiana spoken of by Strabo in the eleventh Booke of his Geography (notwithstanding Cleomedes is of opinion that a perpendicular drawne from the top of the highest hill to the

bottom of the deepest Sea exceeds not this measure), the proportion will be no greater than of one to 1,908. Or let us extend it yet further if you will to foure miles, or thirty-two furlongs (of which height the mountaine Casius, in Syria, is reported by Pliny to be), the proportion will yet be somewhat lesse then of one to 1,789. I am therefore so farre from giving any credit to Patricius, his relations of Tenariffes being seventy-two miles high (unlesse it be measured by many oblique and crooked turnings and windings, in which manner Pliny measureth the height of the Alpes also to be fiftie miles), so that I cannot assent to Alhazan, an Arabian, who would have the toppes of the highest hilles to reach to eight Arabian miles, or eighty furlongs, as I thinke; neither yet to Pliny, who, in his quarto lib., cap. ii, affirmes the mountaine Hæmus to be six miles in height, and I can scarcely yield to the same Pliny when as he speaks of other hilles foure miles in height. And whoever should affirme any hill to be higher than this, though it were Mercury himselfe, I should hardly believe him. Thus much of the height of hilles which seemed to derogate from the roundnesse of the Terrestriall Globe. Patricius proceeds, and goes about to prove that the water also is not round or sphæricall. And he borroweth his argument from the observations of those that conveigh or levell waters, who find by their Dioptricall Instruments that waters have all an equall and plaine superficies, except they be troubled by the violence of windes. On the contrary side, Eratosthenes, in Strabo, affirmes that the superficies of the Sea is in some places higher then it is in other. And he also produceth as assertors of his ignorance those Water-levellers, who, being employed by Demetrius about the cutting away of the Isthmus, or necke of land betwixt Peloponessus and Greece, returned him answere that they found by their Instruments that that part of the Sea which was on Corinth's side was higher than it was at Cenchrææ. The like is also storied of

Lib. 2, c. 65.

L. de Crepuse.

Sesostris, one of the kings of Egypt, who, going about to make a passage out of the Mediterranean into the Arabian Gulfe, is said to have desisted from his purpose because he found that the superficies of the Arabian Gulfe was higher than was the Mediterranean, as it is reported by Aristotle in the end of his first booke of Meteors. The like is also said in the same place by the same Authour to have happened afterward to Darius. Now whether the Architects or Water-levellers employed by Demetrius, Sesostris, and Darius deserve more credit than those whom Patricius nameth I shall not much trouble my selfe to examine. Yet Strabo inveigheth against Eratosthenes for attributing any such eminences and depressions to the superficies of the Sea. And Archimedes his doctrine is that every humid body standing still and without disturbance hath a sphæricall superficies whose centre is the same with that of the Earth. So that wee have just cause to regret the opinions, both of those that contend that the superficies of the Sea is plaine, as also of those that will have it to be in some places higher than in other. Although wee cannot in reason but confesse that so small a portion of the whole Terrestriall Globe as may be comprehended within the reach of our sight, cannot be distinguished by the helpe of any Instruments from a plaine superficies. So that we may conclude Patricius his argument, which he alleadgeth from the experience of Water-conveighers, to be of no weight at all.

Aristotle.

But hee goes on and labours to prove his assertion from the elevation and depression, rising and setting of the Poles and Starres, which were observed daily by those that traverse the Seas ; all which he saith may come to passe, although the surface of the water were plaine. For if any Starre be observed that is in the verticall point of any place, which way soever you travell from that place, the same Starre will seeme to be depressed, and abate something of its elevation, though it were on a plaine superficies. But there is something more in it than Patricius takes notice of. For if wee goe an equall measure of miles, either toward the North or toward the South, the elevation or depression of the Starre will always bee found to be equall: which that it can possibly bee so in a plaine superficies is more than hee will ever be able to demonstrate. If wee take any Starre situate neare the Æquator, the same, when you have removed thence 60 English miles, will be elevated about a degree higher above the Horizón, whether the Starre be directly over your head, or whether you depart thence that so it may bee depressed from your Zenith for 30 or 50 or any other number of degrees. Which that it cannot thus be on a plaine superficies may bee demonstrated out of the principles of Geometry. But yet methinks this one thing might have persuaded Patricius (being so well versed in the Histories of the Spanish Navigations, as his writings sufficiently testifie) that the superficies of the Sea is not plaine, because that the Ship called the *Victory*, wherein Ferdinand Magellane, losing from Spaine and directing his course toward the South-west parts, passed through the Straits, called since by his name, and so touching upon the Cape of Good Hope, having encompassed the whole world about, returned again into Spaine. And here I shall not need to mention the famous voyages of our owne countriemen, Sir Francis Drake and Master Thomas Candish, not so well knowne perhaps abroad, which yet convince Patricius of the same errour. And thus have we lightly touched the chiefe foundations that his cause is built upon; but as for those ill-understood experiments which he brings for the confirmation of the same, I shall let them passe, for that they seeme rather to subvert his opinion than confirme it.

Thus, having proved the Globe of the Earth to be of a Sphericall figure, seeing that the eminency of the highest hills hath scarcely the same proportion to the semidiameter of the Earth that there is betwixt 1 and 1,000, which how small it is any one may easily perceive; I hold it very superfluous to goe about to prove that a Globe is of a figure most proper and apt to expresse the fashion of the Heavens and Earth as being most agreeable to nature, easiest to be understood, and also very beautifull to behold.

Now in Materiall Globes, besides the true and exact description of places, which is, indeed, the chiefest matter to be considered, there are two things especially required. The first whereof is the magnitude and capacity of them, that so there may be convenient space for the description of each particular place or region. The second is the lightnesse of them, that so their weight be not cumbersome. Strabo, in his eleventh booke, would have a Globe to have tenne foot in Diameter, that so it might in some reasonable manner admit the description of particular places. But this bulke is too vast to bee conveniently dealt withall. And in this regard I think that these Globes, of which I intend to speak in this ensuing discourse, may justly bee preferred before all other that have been set before them, as beinge more capacious than any other; for they are in Diameter two foot and two inches, whereas Mercator's Globes (which are bigger than any other ever set before him) are scarcely sixteene inches Diameter. The proportion therefore of the superficies of these Globes to Mercator's will be as 1 to $2\frac{3}{8}$, and somewhat more. Every country, therefore, in these Globes will be above twice as large as it is in Mercator's, so that each particular place may the more easily bee described. And this I would have to bee understood of those great Globes made by William Saunderson of London; concerning the use of which especially we have written this discourse. For he hath set forth other smaller Globes, also, which as they are of a lesser bulke and magnitude, so are they of a cheaper price, that so the meaner Students might herein also be provided for. Now concerning the geographicall part of them, seeing that it is taken out of the newest Charts and descriptions; I am bold to think them more perfect than any other: however they want not their errours. And I thinke it may bee the authors glory to have performed thus much in the edition of these Globes. One thing by the way you are to take notice of, which is that the descriptions of particular places are to be sought for elsewhere, for this is not to be expected in a Globe. And for these descriptions of particular countries, you may have recourse to the Geographicall Tables of Abrahamus Ortelius,[1] whose diligence and industry in this regard seemes to exceed all other before him. To him, therefore, we referre you.[2]

[1] In the edition of 1659 the name of Gerardus Mercator is substituted for that of Abrahamus Ortelius.

[2] In the Dutch editions here follows a long note by Pontanus, describing the globe of Tycho Brahe at Prague, and those of the Duke of Tuscany; and giving the definitions of Euclid.

THE FIRST PART.

Of those things which are common both to the Cœlestiall and Terrestriall Globe.

CHAPTER I.

What a Globe is, with the parts thereof, and of the Circles of the Globe.

A Globe, in relation to our present purpose, we define to be an Analogicall representation either of the Heavens or the Earth. And we call it Analogicall, not only in regard of its forme expressing the Sphæricall figure as well of the Heavens, as also of the Terrestriall Globe, consisting of the Earth itselfe, together with the interflowing Seas; but rather because that it representeth unto us in a just proportion and distance each particular constellation in the Heavens, and every severall region and tract of ground in the Earth; together with certaine circles, both greater and lesser, invented by Artificers for the more ready computation of the same. The greater Circles we call those which divide the whole superficies of the Globe into two equall parts or halves; and those the lesser which divide the same into two unequall parts.[1] Globus quid.

Besides the body of the Globe itselfe, and those things which we have said to be thereon inscribed, there is also annexed a certain frame with necessary instruments thereto belonging, which we shall declare in order.

[1] Here Pontanus inserts another long note, in the Dutch edition, respecting a discussion between Tycho Braye and Peter Ramus, on the method of astronomical computation in use among the ancient Egyptians.

The fabricke of the frame is thus: First of all there is a Base, or foot to rest upon, on which there are raised perpendicularly sixe Columnes or Pillars of equall length and distance; upon the top of which there is fastened to a levell and parallel to the Base a round plate or circle of wood, of a sufficient breadth and thicknesse, which they call the Horizon, because that the uppermost superficies thereof performeth the office of the true Horizon. For it is so placed that it divideth the whole Globe into two equall parts. Whereof that which is uppermost representeth unto us the visible Hemisphære, and the other that which is hid from us. So likewise that Circle which divides that part of the world which wee see from that other which wee see not, is called the Horizon. And that point which is directly over our heads in our Hemisphære, and is on every side equidistant from the Horizon, is commonly called Zenith; but the Arabians name it Semith. But yet the former corrupted name hath prevailed, so that it is always used among Writers generally. And that point which is opposite to it in the lower Hemisphære the Arabians call Nathir; but it is commonly written Nadir. These two points are called also the Poles of the Horizon.

Horizon.

Furthermore, upon the superficies of the Horizon in a Materiall Globe, there are described, first, the twelve Signes of the Zodiaque, and each of these is again divided into thirty lesser portions; so that the whole Horizon is divided into 360 parts, which they also call degrees. And if every degree be divided into sixtie parts also, each of them is then called a Scruple or Minute; and so by the like subdivision of minutes into sixtie parts will arise Seconds, and of these Thirds, and likewise Fourths and Fifths, etc., by the like partition still of each into sixtie parts.[1]

There is also described upon the Horizon the Roman

[1] Pontanus adds, in a note, that the days of the month, and the Roman Kalends, Nones, and Ides, are also marked on the modern horizon.

Calendar, and that three severall ways; to wit, the ancient way, which is still in use with us here in England; and the new way appointed by Pope Gregory 13, wherein the Equinoxes and Solstices were restored to the same places wherein they were at the time of the celebration of the Councell of Nice; in the third, the said Equinoctiall and Solsticiall points are restored to the places that they were in at the time of our Saviour Christ's nativity. The months in the Calendar are divided into dayes and weekes, to which are annexed, as their peculiar characters, the seven first letters of the Latine Alphabet. Which manner of designing the dayes of the Moneth was first brought in by Dionysius Exiguus, a Romane Abbot, after the Councell of Nice.

The innermost border of the Horizon is divided into 32 parts, according to the number of the Windes, which are observed by our moderne Sea-faring men in their Navigations; by which also they are wont to designe forth the quarters of the Heavens and the coasts of Countries. For the Ancients observed but foure winds only, to which were after added foure more; but after ages, not content with this number, increased it to twelve, and at length they brought it to twenty-foure, as Vitruvius notes. And now these later times have made them up thirty-two, the names whereof both in English and Latine are set down in the Horizon of Materiall Globes.[1]

There is also let into this Horizon two notches opposite one to the other, a circle of brasse, making right angles with the said Horizon, and placed so that it may be moved at pleasure both up and downe by those notches, as neede shall require. This Circle is called the Meridian, because that one side of it, which is in like manner divided into 360 degrees, supplyeth the office of the true Meridian. Now the meridian is one of the greater circles passing through the Poles of the World and also of the Horizon; to which, when Meridianus.

[1] Pontanus here inserts a note on the uses of the horizon.

the Sunne in his daily revolution is arrived in the upper Hemisphære, it is midday; and when it toucheth the same in the lower Hemisphære it is midnight at that place whose Meridian it is.

These two Circles, the Horizon and Meridian, are various and mutable in the Heavens and Earth, according as the place is changed. But in the Materiall Globe they are made fixed and constant; and the earth is made moveable, that so the Meridian may be applied to the Verticall point of any place.[1]

Poli. Boreus and Austrinus.

In two opposite poynts of this Meridian are fastened the two ends of an iron pinne passing through the body of the Globe and its center. One of which ends is called the Arcticke or North Pole of the World; and the other the Antarcticke or South Pole; and the pinne itselfe is called the Axis. For the Axis of the World is the Diameter about which it is turned; and the extreme ends of the Axis are called the Poles.

Horarius.

To either of these Poles, when need shall require, there is a certain brasse circle or ring of a reasonable strong making to be fastened, which circle is divided into 24 equall parts, according to the number of the houres of the day and night; and it is therefore called the Houre circle. And this circle is to be applied to either of the Poles in such sort as that the Section where 12 is described may precisely agree with the points of mid-day and mid-night in the superficies of the true Meridian.

Index Horarius.

There is also another little pinne or stile to be fastened to the end of the Axis, and in the very center of the Houre circle; and this pinne is called in Latine, Index Horarius, and so made as that it turnes about and pointeth to every of the 24 sections in the Houre Circle, according as the Globe it selfe is moved about; so that you may place the point of it to what houre you please.[2]

1 Pontanus here has a note on the uses of the meridian.

2 Here Pontanus has a note on using the hour circle, meridian, and quadrant of altitude.

CHAPTER II.

Of the Circles which are described upon the Superficies of the Globe.

And now in the next place we will shew what Circles are described upon the Globe it selfe. And first of all there is drawne a circle in an equall distance from both the Poles, that is 90 degrees, which is called the Æquinoctiall or Equator; because that when the Sunne is in this Circle days and nights are of equall length in all places. By the revolution of Circle is defined a naturall day, which the Greeks call *νυχθημερον*. For a day is twofold: Naturall and Artificiall. A Naturall day is defined to be the space of time wherein the whole Æquator makes a full revolution; and this is done in 24 houres. An Artificiall day is the space wherein the Sunne is passing through our upper Hemisphære; to which is opposed the Artificiall night, while the Sunne is carried about in the lower Hemisphære. So that an Artificiall day and night are comprehended within a Naturall day.

Æquator.

Dies Naturalis: Artificialis.

The Parts of a day are called houres; which are either equall or unequall. An Equall houre is the 24th part of a Naturall day, in which space 15 degrees of the Æquator doe always rise, and as many are depressed on the opposite part. An Unequall houre is the 12th part of an Artificiall day, betwixt the time of the Suns rising and setting againe. These Houres are againe divided into Minutes. Now a minute is the 60th part of an houre; in which space of time a quarter of a degree in the Æquator, that is 15 minutes, doe rise and as many set.[1]

Horæ æquales.

Inæquales.

The Æquator is crossed or cut in two opposite points by an oblique Circle, which is called the Zodiack. The obliquity of this Circle is said to have beene first observed by

Zodiacus.

[1] Here Pontanus has a note on the uses of the equator.

Anaximander Milesius, in the 58 Olympiad, as Pliny writeth in his lib. 2, cap. 8. Who also in the same place affirmes that it was first divided into 12 parts which they call Signes by Cleostratus Tenedius, in like manner as we see it at this day. Each of these Signes is again subdivided into 30 Parts, so that the whole Zodiack is divided in all into 360 parts, like as the other circles are. The first twelfth part whereof, beginning at the Vernall Intersection, where the Æquator and Zodiack crosse each other, is assigned to Aries, the second to Taurus, etc., reckoning from West to East. But here a young beginner in Astronomy may justly doubt what is the reason that the first 30 degrees or 12th part of the Zodiack is attributed to Aries, whereas the first Starre of Aries falls short of the Intersection of the Æquinoctiall and Zodiacke no less than 27 degrees. The reason of this is because that in the time of the Ancient Greeks, who first of all observed the places and situation of the fixed Starres and expressed the same by Asterismes and Constellations, the first Starre of Aries was then a very small space distant from the very Intersection. For in Thales Milesius his time it was two degrees before the Intersection; in the time of Meton the Athenian, it was in the very Intersection. In Timocharis his time it came two degrees after the Intersection. And so by reason of its vicinity the Ancients assigned the first part of the Zodiack to Aries, the second to Taurus, and so the rest in their order; as it is observed by succeeding ages even to this very day.[1]

Under this Circle the Sunne and the rest of the Planets finish their severall courses and periods in their severall manner and time. The Sunne keepes his course in the middest of the Zodiack, and therewith describeth the Ecliptick circle. But the rest have all of them their latitude and deviations from the Suns course or Ecliptick. By reason of which their digressions and extravagancies the

[1] Pontanus here gives a note on Thales and Meton.

Ancients assigned the Zodiaque 12 degrees of latitude. But our moderne Astronomers, by reason of the Evagations of Mars and Venus, have added on each side two degrees more; so that the whole latitude of the Zodiack is confined within 16 degrees. But the Ecliptick onely is described on the Globe, and is divided in like manner as the other Circles into 360 degrees.[1]

The Sunne runneth thorough this Circle in his yearly motion, finishing every day in the yeare almost a degree by his Meane motion, that is 59 min. 8 seconds. And in this space he twice crosseth the Æquator in two poynts equally distant from each other. So that when he passeth over the Æquator at the beginnings of Aries and Libra, the dayes and nights are then of equall length. And so likewise when the Sunne is now at the farthest distance from the Æquator, and is gotten to the beginning of Cancer or Capricorne, he then causeth the Winter and Summer Solstices. I am not ignorant that Vitruvius, Pliny, Theon Alexandrinus, Censorinus, and Columella, are of another opinion (but they are upon another ground); when as they say that the Æquinoxes are, when as the Sunne passeth through the eighth degree of Aries and Libra, and then it was the midst of Summer and Winter, when the Sun entered the same degree of Cancer and Capricorne. But all these authors defined the Solstices by the returning of the shadow of dials: which shadow cannot bee perceived to returne backe againe, as Theon saith, till the Sunne is entered into the eighth degree of Libra and Aries.[2] C. 2, frag. quod Censor adiungitur.

The Space wherein the Sunne is finishing his course through the Zodiack is defined to be a Yeare, which consists of 365 dayes, and almost 6 houres. But they that think to find the exact measure of this period will find themselves frustrate; for it is finished in an unequall time. It hath beene alwayes a controversie very much agitated among the Annus.

[1] Pontanus here has a note on the ecliptic and zodiac.

[2] Here Pontanus inserts a note on the uses of the zodiac.

Ancient Astronomers, and not yet determined. Philolaus, a Pythagorean, determines it to be 365 dayes; but all the rest have added something more to this number. Harpalus would have it to be 365 dayes and a halfe; Democritus 365 dayes and a quarter, adding beside the 164 part of a day. Œnopides would have it to be 365 dayes 6 houres, and almost 9 houres. Meton the Athenian determined it to be 365 dayes, 6 houres and almost 19 minutes. After him Calippus reduced it to 365 dayes and 6 houres, which account of his was followed by Aristarchus of Samos, and Archimedes of Syracusa. And according to this determination of theirs Julius Cesar defined the measure of his Civile year, having first consulted (as the report goes) with one Sosigenes, a Peripateticke and a great Mathematician. But all these, except Philolaus (who came short of the just measure), assigned too much to the quantity of a yeare. For that it is somewhat lesse than 365 dayes 6 houres is a truth confirmed by the most accurate observations of all times, and the skilfullest artists in Astronomicall affaires. But how much this space exceedeth the just quantity of a yeare is not so easy a matter to determine. Hipparchus, and after him Ptolomy, would have the 300 part of a day subtracted from this measure (for Jacobus Christmannus was mistaken when he affirmed that a Tropicall yeare, according to the opinions of Hipparchus and Ptolomy, did consist of 365 dayes and the 300 part of a day). For they doe not say so, but that the just quantity of a yeare is 365 dayes and 6 houres, abating the 300 part of a day, as may be plainely gathered out of Ptolomy, *Almagest.*, lib. 3, cap. 2, and as Christmannus himselfe hath elsewhere rightly observed. Now, Ptolomy would have this to be the just quantity of a yeare perpetually and immutably; neither would he be perswaded to the contrary, notwithstanding the observations of Hipparchus concerning the inequallity of the Sunnes periodicall revolution. But yet the observations of succeeding times, compared with those of Hipparchus and

Joseph. Scal. de Em. temp.

Ad C. 15. Alfrag.

Ptolomy, doe evince the contrary. The Indians and Jewes subtract the 120 part of a day; Albategnius, the 600 part; the Persians, the 115 part, according to whose account Messahalah and Albumazar wrote their tables of the Meane Motion of the Sunne. Azaphius Avarius and Arzachel affirmed that the quantity assigned was too much by the 136 part of a day; Alphonsus abateth the 122 part of a day; some others, the 128 part of a day; and some, the 130 part of a day. Those that were lately employed in the restitution of the Romane Calendar would have almost the 133 part of a day to be subtracted, which they conceived in 400 years would come to three whole dayes. But Copernicus observed that this quantity fell short by the 115 part of a day. Most true therefore was that conclusion of Censorinus, that a yeare consisted of 365 dayes, and I know not what certaine portion, not yet discovered by Astrologers. Censo. c. 21.

By these divers opinions here alledged is manifestly discovered the error of Dion, which is indeed a very ridiculous one. For he had conceit that in the space of 1461 Julian yeares there would be wanting a whole day for the just measure of a yeare; which he would have to be intercaled, and so the Civile Julian Yeare would accurately agree with the revolution of the Sunne. And Galen also, the Prince of Physitians, was grossly deceived when he thought that the yeare consisted of 365 dayes 6 houres, and besides almost the 100 part of a day; so that at every hundred yeares end there must be a new intercalation of a whole day. Dion, l. 43. L. 4, c. 3. Progn.

Now, because the Julian yeare (which was instituted by Julius Cæsar, and afterwards received and is still in use) was somewhat longer than it ought to have beene, hence it is that the Æquinoxes and Solstices have gotten before their Ancient situation in the Calendar. For about 432 yeares before the incarnation of our Saviour Christ, the Vernall Æquinoxe was observed by Meton and Euctemon to fall on the 8 of the Kalends of Aprill, which is the 25 of March Æquinoc. et solatis mutatio.

according to the Computation of the Julian Yeare. In the yeare 146 before Christ it appeares, by the observation of Hipparchus, that it is to be placed on the 24 of the same moneth, that is the 9 of the Kalends of Aprill. So that from hence we may observe the error of Sosigenes (notwithstanding he was a great Mathematician), in that above 100 yeares after Hipparchus, in instituting the Julian Calendar, he assigned the Æquinoxes to be on the 25 of March or the 8 of the Kalends of Aprill, which is the place it ought to have had almost 400 years before his time. This error of Sosigenes was derived to succeeding ages also; insomuch that in Galens time, which was almost 200 yeares after Julius Cæsar, the Æquinoxes were wont to be placed on the 24 day of March and September, as Theodorus Gaza reports. In the yeare of our Saviours Incarnation it happened on the 10 of the Kalends of April or the 23 of March. And 140 years after, Ptolomy observed it to fall on the 11 of the Kalends. And in the time of the Councell of Nice, about the yeare of our Lord 328, it was found to be on the 21 of March, or the 12 of the Kalends of Aprill. In the yeare 831 Thebit Ben Chorah observed the Vernall Æquinoxe to fall on the 17 day of March: in Alfraganus his time it came to the 16 of March. Arzachel, a Spaniard, in the yeare 1090, observed to fall on the Ides of March, that is the 15 day. In the yeare 1316 it was observed to be on 13 day of March. And in our times it has come to be on the 11 and 10 of the same moneth. So that in the space of 1020 yeares, or thereabout, the Æquinoctiall points are gotten forward no lesse then 14 dayes. The time of the Solstice also, about 388 yeares before Christ, was observed by Meton and Euctemon to fall upon the 18 day of June, as Joseph Scaliger and Jacobus Christmannus have observed. But the same in our time is found to be on the 12 of the same moneth.

Gaza de Mens. Attic.

The Eclipticke and Æquator are crossed by two great Circles also, which are called Colures; both which are drawne through the Poles of the world, and cut the Æquator at right Angles. The one of them passing through the points of both the Intersections, and is called the Eqinoctiall Colure; the other passing through the points of the greatest distance of the Zodiack from the Æquator, is therefore called the Solsticiall Colure.[1]

Coluri Solstitiorum et æquinoctiorum.

Now that both the colures, as also the Æquinoctiall points have left the places where they were anciently found to be in the Heavens, is a matter agreed upon by all those that have applyed themselves to the observations of the Cœlestiall motions; only the doubt is whether fixed Starres have gone forward unto the preceding Signes, as Ptolomy would have it, or else whether the Æquinoctiall and Solsticiall points have gone back to the subsequent Signes, according to the Series of the Zodiack, as Copernicus opinion is.[2]

The first Starre of Aries, which in the time of Meton the Athenian, was in the very Vernall Intersection, in the time of Thales Milesius was two degrees before the Intersection. The same in Timochares his time, was behind it two degrees 24 minutes; in Hipparchus time, 4 degrees 40 minutes; in Albumazars time, 17 degrees 50 minutes; in Albarenus his time, 18 degrees 10 minutes; in Arzachels time, 19 gr. 37 minutes; in Alphonsus his time, 23 degrees 48 minutes; in Copernicus and Rhœticus his time, 27 degrees 21 minutes. Whence Franciscus Baroccius is convinced of manifest error in that he affirmes that the first Starre of Aries, at the time of our Saviours Nativity, was in the very Vernall Intersection, especially contending to prove it, as he doth, out of Ptolomy's observations, out of which it plainly appears that it was behind in no lesse then 5 degrees.

Stellarum fixerum longitutidinis Mutatæ.

In Heronis Geodesiam.

In like manner the places of the Solstices are also changed, as being alwayes equally distant from the Æquinoctiall

[1] Pontanus here inserts a note on the office of the colures.

[2] Pontanus, in a long note, here gives the opinions of Scaliger and Tycho Brahe on the precession of the equinoxes.

points. This motion is finished upon the Poles of the Ecliptick, as is agreed upon both by Hipparchus and Ptolomy, and all the rest that have come after them. Which is the reason that the fixed Starres have always kept the same latitude though they have changed their declination. For confirmation whereof many testimonies may be brought out of Ptolomy, lib. 7, cap. 3 Almag. I will only alleadge one more notable then the rest out of Ptolomies Geogr. lib. 1, cap. 7. The Starre which we call the Polar Starre, and is the last in the taile of the Beare, is certainely knowne in our time to be scarce three degrees distant from the Pole, which very Starre in Hipparchus his time was above 12 degrees distant from the Pole, as Marinus in Ptolomy affirmes. I will produce the whole passage which is thus. In the Torrid Zone (saith he) the whole Zodiacke passeth over it, and therefore the shadowes are cast both wayes, and all Starres there are seen to rise and set. Onely the little Beare begins to appeare above the Horizon in those places that are 500 furlongs northward from Ocele. For the Parallel that passeth through Ocele is distant from the Æquator 11 gra. $\frac{2}{5}$. And Hipparchus affirmes that the Starre in the end of the little Beares taile, which is the most Southward of that Constellation, is distant from the Pole 12 gr. $\frac{2}{5}$. This excellent testimony of his, the Interpreters have, in their translating, the place most strangely corrupted (as both Johannes Wernerus and after him P. Nonius have observed), setting down instead of 500 Quinque Mille 5000, and for Australissimam, the most Southerne, Borealissimam, the most Northerly: being led into this error perhaps, because that this Starre is indeed in our times the most Northerly. But if these testimonies of Marinus and Ptolomy in this point be suspected, Strabo in his lib. 2, Geogr., shall acquit them of this crime. And he writes thus. It is affirmed by Hipparchus (saith he) that those that inhabit under the Parallel that runneth thorough the Coun-

Mutata declivat, Stell. fixarum.

Strabo.

trey called Cinnamomifera (which is distant from Meroe, Southward 3000 furlongs, and from the Æquinoctiall 8800), are situated almost in the midst betwixt the Æquator and the Summer Tropicke, which passeth through Syene (which is distant from Meroe 5000 furlongs), and these that dwell here are the first that have the Constellation of the little Beare inclosed within their Arcticke Circle, so that it never sets with them, for the bright Starre that is seen in the end of the taile (which is also the most Southward of all) is so placed in the very Circle itselfe, that it doth touch the Horizon. This is the testimony of Strabo, which is the very same that Ptolomy and Marinus affirme, saving that both in this place and elsewhere he alwayes assignes 700 furlongs in the Earth to a degree in the Heavens, according to the doctrine of Eratosthenes, whereas both Marinus and Ptolomy allow but 500 onely ; of which we shall speak more hereafter.

Let us now come to the lesser circles which are described in the Globe. And these are all parallel to the Equator ; as first of all the Tropickes, which are Circles drawn through the points of the greatest declination of the Eclipticke on each side of the Æquator. Of which, that which looks toward the North Pole is called the Tropicke of Cancer ; and the other, bordering on the South, the Tropicke of Capricorne. For the Sunne in his yearely motion through the Eclipticke arriveing at these points, as his utmost bounds, returneth againe toward the Æquator. This Retrocession is called by the Greeks τροπη, and the parallel circles drawne through the same points are likewise called Tropickes.[1]

Tropici Canceri et Capricorni.

The distance of the Tropickes from the Æquator is diversely altered, as it may plainely appear, by comparing the observations of later times with these of the Ancients. For not to speake anything of Strabo, Proclus, and Leontius Mechanicus, who all assigned the distance of either Tropicke from the Æquator to be 24 degrees (for these seeme to have

Mutatis Solis declinatio Mun.

[1] Pontanus here adds a note on the uses of the tropics.

handled the matter but carelessly) we may observe the same from the more accurate observations of the greatest Artists. For Ptolomy found the distance of either Tropicke to be 23 gr. 51 min. and 1/5 just as great as Eratosthenes and Hipparchus had found it before him; and therefore he conceived it to be immutable. Machomethes Aratensis observed this distance to be 23 degrees 35 minutes, right as Almamon, King of Arabia, had done before him. Arzahel, the Spaniard, found it to be in his time 23 degrees 34 minutes; Almeḥon the Sonne of Almuhazar, 23 degrees 33 minutes and halfe a minute; Prophatius, a Jew, 23 degrees 32 minutes; Purbachius and Regiomontanus, 25 degrees 28 minutes; Johan Wernerus, 23 degrees 28 minutes and an halfe; and Copernicus found it in his time to be just as much.[1]

There are two other lesser circles described in an equall distance from the Poles to that of the Tropickes from the Æquator, which circles take their denomination from the Pole on which they border. So that one of them is called the Arcticke or North Circle, and the opposite Circle the Antarcticke or Southerne. In these Circles the Poles of the Eclipticke are fixed, the Solsticiall Colure crossing them in the same place. Strabo, Proclus, Cleomedes, all Greeke Authors, and some of the Latines also, assigne no certaine distance to these circles from the Poles; but make them various and mutable, according to the diversity of the elevation of the Pole or diverse position of the Sphære; so that one of them must be conceived to be described round about that Pole which is elevated, and to touch the very Horizon, and is therefore the greatest of all the parallels that are always in sight; and the other must be imagined as drawne in an equall distance from the opposite Pole; and this is the greatest of those parallels that are always hidden.

Circuli Arct. et Antarct.

[1] Pontanus here inserts a table of the distances of the tropics from the equator, at various epochs, as calculated by the astronomers mentioned in the text, adding remarks by Tycho Brahe on the subject.

Besides the circles expressed in the Globe there are also some certaine other circles in familiar use with the Practicall Astronomers, which they call verticall circles. These are greater circles drawne from the verticall point through the Horizon, in what number you please; and they are called by the Arabians Azimuth, which appellation is also in common use among our Astronomers. The Office of these circles is supplied by the helpe of a quadrant of Altitude, which is a thin plate of brasse divided into 90 degrees. This quadrant must bee applied to the vertex of any place when you desire to use it, so that the lowest end of it, noted with the number of 90, may just touch the horizon in every place. The quadrant is made moveable, that so it may be fastened to the verticall point of any place.

Circuli Vesticules.

Quadrans Altitudin.

CHAPTER III.

Of the three positions of Sphœres: Right, Parallel, and Oblique.

According to the diverse habitude of the Æquator to the Horizon (which is either parallel to it, or cutteth it, and that either in oblique or else in right angles) there is a threefold position or situation of Spheres. The first is of those that have either Pole for their verticall point, for with these the Æquator and Horizon are Parallel to each other, or indeed rather make but one circle betwixt them both. The 2d is of those whose Zenith is under the Æquator. The third agreeth to all other places else. The first of these situations is called a Parallel Sphere; the second, a Right; and the third an Oblique Sphere. Of these severall kindes of position the two first are simple, but the third is manifold and divers, according to the diversity of the latitude of places. Each of these have their peculiar properties.

Triplex Sphære positio.

Spheræ parallela recta, obliqua.

Sph. Parall. accidentia.

Those that inhabite in a Parallel Sphære see not the Sun or other Stars either rising or setting, or higher or lower, in the diurnall revolution. Besides, seeing that the Sun in his yearely motion traverseth the Zodiaque which is divided by the Æquator into 2 equall parts; one whereof lieth toward the North, and the other toward the South; by this means it comes to passe, that while the sun is in his course through those figures that are nearest the Verticall Pole, all this while hee never setteth, and so maketh but one continued artificiall day, which is about the space of sixe moneths. And so contrariwise, while he runneth over the other remoter figures lying toward the Opposite Pole, hee maketh a long continuall night of the like space of time or thereabout. Now at such time as the Sun in his diurnall revolution shall come to touch the very Æquator, he is carried about in such sort as that he is not wholly apparent above the Horizon, nor yet wholly hidden under it, but as it were halfe cut off.

Affectiones. Sphæ. Rect.

The affections of a Right Sphære are these. All the Stars are observed to rise and set in an equall space of time, and continue as long above the Horizon as they doe under it. So that the day and night here is always of equall length.[1]

Sphæ. Oblique conuenient.

An Oblique Sphære hath these properties. Their dayes sometimes are longer then their nigl ts, sometimes shorter, and sometimes of equall length. For when the Sun is placed in the Æquinoctiall points, which (as wee have said) happeneth twice in the yeare, the daies and nights are then equall. But as he draweth nearer to the elevated Pole the dayes are observed to increase and the nights to decrease, till such time as hee comes to the Tropique, when as he there maketh the longest dayes and the shortest nights in the yeare. But when he returneth toward the Opposite Pole

[1] Pontanus, in a note, doubts whether this does not agree with the rational or intelligible rather than with the sensible horizon: because, even in a right sphere, the sight can hardly reach both the Poles, by reason of the exuberancy of the earth.

the dayes then decrease till he toucheth the Tropique that lieth nearer the same Pole, at which time the nights are at the longest and the dayes shortest. In this position of Sphære also some Starres are never seene to set; such as are all those that lie within the compasse of a Circle described about the Elevated Pole and touching the Horizon; and some in like manner are never observed to appeare above the Horizon; and these are all such Starres as are circumscribed within the like Circle drawne about the Opposite Pole. These Parallel Circles (as wee have said) are those which the Greekes, and some of the Latines also, call the Arctique and Antarctique Circles, the one alwayes appearing and the other always lying hid. All the other Starres which are not comprehended within these two Circles have their rising and settings by course. Of which those that are placed between the Æquator and this always apparent Circle, continue a longer space in the upper Hemisphære and a lesse while in the lower. So, on the contrary, those that are nearer to the Opposite Circle are longer under the Horizon, and the lesse while above it. Of all which affection this is the cause. The Sunne being placed in the Æquator (or any other Starre) in his daily revolution describeth the Æquinoctiall circle; but being without the Æquator he describeth a greater or lesser Parallel, according to the diversity of his declination from the Æquator. All which Parallels, together with the Æquator itselfe, are cut by the Horizon in a Right Sphære to right angles. For when the Poles lie both in the very Horizon, and the Zenith in the Æquator, it must needs follow that the Horizon must cut the Æquator in right angles, because it passeth through its Poles. Now, because it cutteth the Æquator at right angles, it must also necessarily cut all other circles that are Parallel to it in right angles; and, therefore, it must needs divide them into two equall parts. So that if halfe of all these Parallels, as also of the Æquator, be above the Horizon, and the other halfe lye hid under it, it must necessarily follow that the Sunne, and other Starres, must be as long in passing through tho Upper Hemisphære as through the lower. And so the daies must be as long as the nights, as all the Starres in like manner will be 12 houres above the Horizon, and so many under it. But in an Oblique Sphære, because one of the Poles is elevated above the Horizon and the other is depressed under it, all things happen cleane otherwise. For seeing that the Horizon doth not passe through the Poles of the Æquator, it will not therefore cut the Parallels in the same manner as it doth the Æquator; but those Parallels that are nearest to the elevated Pole will have the greatest portion of them above the Horizon and the least under. But those that are nearest the opposite Pole will have the least part of them seene, and the greatest part hid; only the Æquator is still divided into two equall parts, so that the conspicuous part is equall to that which is not seene. And hence it is that in all kinds of Obliquitie of Sphære, when the Sun is in the Æquator, the day and night is alwayes of equall length. And as he approacheth towards the elevated Pole the dayes encrease; because the greater Arch or portion of the Parallels is seene. But when he is nearer the hidden Pole the nights are then the longest, because the greatest segment of those Parallels are under the Horizon. And by how much Higher either Pole is elevated above the Horizon of any Place, by so much the dayes are the longer in Summer and the nights in Winter.[1]

[1] Pontanus here explains the errors of Clavius and Sacrobosco respecting the spheres, while expressing concurrence with our author.

CHAPTER IIII.

Of the Zones.

The foure lesser Circles which are Parallel to the Æquator divide the whole Earth into 5 partes, called, by the Greekes, Zones. Which appellation hath also beene received and is still in use among our Latine Writers; notwithstanding they sometimes also use the Latine word, *Plaga*, in the same signification. But the Greekes do sometimes apply the word Zona to the Orbes of the Planets (in a different sense than is ever used by our Authors), as may appear by that passage of Theon Alexandrinus in his commentaries upon Aratus —*ἔχει γὰρ ὁ ουρανὸς ζ ζωνας ὀυκ επιψαυγσας τω ζωδιακω ων τιω μ πρωτιω εχει ὁ κρονος τω δε δευτεραν ο Ζευς*;[1] that is: There are also in the Heavens seven Zones which are not contiguous to the Zodiaque; the first whereof is assigned to Saturne, the second to Jupiter, etc.

Of these five Zones three were accounted by the Ancient Philosophers and Geographers to bee inhabitable and intemperate. One of them by reason of the Sunnes beames continually beating upon the same, and this they called the Torrid Zone, and is terminated by the Tropiques on each side, and the other two, by reason of extreme cold, they thought could not be inhabited, as being so remote from the heat of the Sunnes beames; whereof one was comprehended within the Arctique Circle, and the other within the Antarctique. But the other two were accounted temperate, and therefore habitable, the one of them lying betwixt the Arctique Circle and the Tropique of Cancer; and the other betwixt the Antarctique and the Tropique of Capricorn.

Zone tres intemperata. Vna Æstu. Dua frigere.

Neither did this opinion (although in a manner generally

[1] ΑΡΑΤΟΥ ΣΟΛΕΩΣ Φαινομενα καὶ Διοσμεια: Θεωνος Σχολια (*Oxonii*, 1672), p. 57.

received among the Ancients) concerning the number and bounds of the Zones, even then want its opposition. For
Strabo, l. 2. Parmenides would have that Zone, which they call the Torrid, to be extended far beyond the Tropiques; so that he made it almost as large againe as it ought to have beene; but is withall reprehended for it by Posidonius, because he knew that above half of that space which is contained betwixt our Summer Tropiques and the Æquator was inhabited. So likewise Aristotle terminated the Torrid Zone betwixt the Tropiques, and the Temperate Zones with the Tropiques and the Arctique and Antarctique Circles. But he is also taxed by the same Posidonius in that he appoints the Arctique Circles, which the Greekes will have to be mutable, to be the limits of the Zones.

Polybius makes five Zones by dividing the Torrid into two parts, and reckoning one of them with the Winter Tropique to the Æquinoctiall, and the other from thence to the Summer Tropique. Others, following Eratosthenes, would have a certaine narrow Zone which should be temperate and fit for habitation under the Æquinoctiall line; of which opinion was Avicen the Arabian. And some of our Moderne Writers (as Nicolaus Lyronus, Thomas Aquinas, and Campanus), I know not upon what grounds, will have the Terrestriall Paradise, spoken of in the beginning of Genesis, to be placed under the Æquinoctiall Line. And so likewise, Eratosthenes and Polybius would have all that which they call the Torrid Zone to be temperate. In like manner
Cleomedes. Posidonius contradicted the received opinion of the Ancient Philosophers, because he knew that both Lyene, which place lieth under the Tropic of Cancer, and also Æthiopia, which lieth more inward, and over whose heads the Sun lieth longer then it doth upon theirs under the Æquator, are notwithstanding inhabited. Whence he concluded that the parts under the Æquinoctiall are not inhabited, because he saw that those under the Tropique wanted not inhabitants. Yet Ptolomy, in his 2d booke and sixe chapter of his Almagest conceiveth

all those things which are reported of the temperatenesse under the line, to be rather conjecture then truth of story; and yet in the last chapter of the fifth booke of his Geography, he describes us a country in Æthiopia which he calleth Agisymba, and placeth farre beyond the Æquinoctiall (notwithstanding some of our Moderne Geographers sticke not to place it Northward from the Æquator contrary to Ptolomies mind). This inconsistency of Ptolomy has given occasion to some to suspect that the Almagest and Cosmography were not the same Author's Works.[1]

Jacob. Chri.

Now, as concerning these conceits of the Ancients about the number of the intemperate Zones, if they were not sufficiently proved to be vaine and idle, by the authority of Eratosthenes and Polybius; yet certainely it is very evidently demonstrated by the Navigations both of the Portugalls, and also of our own Countrymen, that not only that tract of land which the Ancients called the Torrid Zone is fully inhabited, but also that within the Arctique Circle, above 70 degrees from the Æquator, all places are full of inhabitants. So that now no man needs to doubt any further of the truth of this; unless he had rather erre with Sacred and Venerable Antiquity, then be better informed by the experience of Moderne Ages, though never so strongly backed with undeniable proofes and testimonies.

CHAPTER V.

Of the Amphiscii, Heteroscii, and Periscii.

Amphiscii.

The inhabitants of these Zones, in respect of the diversity of their noon shadowes, are divided into three kinds, Amphiscii, Heteroscii, and Periscii. Those that inhabite betwixt the two Tropics are called Amphiscii, because that their noon shadowes are diversely cast, sometime toward the

[1] Pontanus here points out similar inconsistencies in Pliny.

South, as when the Sunne is more Northward then their Verticall point, and sometimes more toward the North, as when the Sun declines Southward from their Zenith.

Heteroscii. Those that live betwixt the Tropiques and Arctique circles, are called Heteroscii, because the shadowes at noone are cast onely one way, and that either North or South. For the Sunne never comes farther North then our Summer Tropick, nor more Southward then the Winter Tropicke. So that those that inhabite Northward of the Summer Tropique have their shadowes cast alwayes toward the North; as in like manner those that dwell more Southward then the Winter Tropick have their Noone Shadowes cast alwayes toward the South. Those that inhabite betwixt the Arctique or Antarctique Circles and the Poles, are called Periscii, because that the Gnomons doe cast their shadowes circularly; and the reason hereof is, for that the Sun is caried round about above their Horizon in his whole Diurnall Revolution.

Periscii.

CHAPTER VI.

Of the Periæci, Antæci, and Antipodes.

The inhabitants of the temperate Zones have by the Ancient Geographers beene divided in respect either of the same Meridian, or Parallel, or else equall situation in respect of divers parts of the Æquator, in such sort as that to every habitation in these severall parts they have added three other different in position whose inhabitants they called Periæci, Antæci, and Antipodes.

Periæci. Periæci are those that live under the same Meridian, and and the same Parallel also, yet equally distant from the Æquator but in two opposite points of the same Parallel.

Antæci. Antæci are such has have the same Meridian, but live in diverse Parallels, yet equally distant from the Æquator though in diverse parts.

Antipodes (which are called Antichthones) are such as inhabite under one Meridian, but under two diverse Parallels, which are equally distant from the Æquator, and in opposite points of the same; or else wee may define them to be such as inhabite two places of the earth, which are Diametrically opposite. Antipodes.

They, therefore, which are Periæci in respect of us, are Antæci in respect to our Antipodes; and those that are Antæci to us are Periæci to our Antipodes, and our Periæci are Antipodes to those which are Antæci to us.

We have also many accidents common to our Periæci. For we both inhabite the same temperate Zone: and have Summer, Winter, increase and decrease of daies and nights at the same time. Only this difference is betwixt us, that when it is noon with us it is midnight with them. Those Authors that have added this difference also, that when the Sun rises with us it setteth with those that are our Periæci, have betrayed their own ignorance. For, if this were so, it would then follow, that, when the day is longest with us, it should be at the shortest with them; but this is most false. They have committed the like errour concerning our Antæci also, when as they will have the Sun to rise with us and them at the same time. The ground of which their errour perhaps may be in that they conceived us and our Antæci to have the same Horizon, but that ours was the uppermost Hemisphære and theirs the lower; the like they conceived of our Periæci. But this is an errour unworthy of those that are but meanely versed in Astronomy. Eorum Comparationes.

We agree with our Antæci in this, that we have midday and midnight both at the same time. But herein we differ that the seasons of the yeare are cleane contrary. For when wee have Summer they have Winter, and our longest day is the shortest with them. We also inhabite temperate Zones both of us, though different from each other in the times and seasons.

But with our Antipodes all things are quite contrary, both dayes and nights with their beginnings and endings, as also the seasons of the Yeare. For at what time we, through the benefit of the Sunne, enjoy our Summer and the longest day, then is it winter with them, and the dayes at the shortest. So likewise when the Sun riseth with us it setteth with them ; and so contrariwise, when it setteth with us it riseth with them. For we inhabite the upper Hemisphære, and they the lower divided by the same Horizon.

CHAPTER VII.

Of Climates and Parallels.

According to the different quantity of the longest dayes, Geographers have divided the whole earth, on each side of the Æquator to the Poles, into Climates and Parallels. A
Clima. Climate they define to be a space of earth comprehended betwixt any two places whose longest dayes differ in quan-
Parallela. tity halfe an houre. And a Parallel is a space wherein the dayes increase in length a quarter of an houre; so that every Climate containeth two Parallels. Those Climates, as also the Parallels themselves, are not all of equall quantity. For the first Clime (as also the Parallel) beginning at the Æquator is larger than the second, and the second is likewise greater than the third. Only herein they all agree that they differ equally in the quantity of the longest day.

The Ancients reckoned but 7 Climates at the first ; to which number were afterward added two more, so that in the first of these numbers were comprehended 14 Parallels, but in the later 18. Ptolomy accointing the Parallels by the difference of a quarter of an houre, reckoneth in all 24; by whole houres difference, 4; by whole moneths, 6. So that besides the Æquator, reckoning the whole number of Parallels on each side, they amount to 38.

In the Meridian of a Materiall Globe there are described nine Climates differing from each other by the quantity of halfe an houre. After these there are other also set according to the difference of an whole houre; and last of all those that differ in whole months are continued to the very Pole, each of them expressed in their severall latitudes.

THE SECOND PART.

CHAPTER I.

Of such things as are proper to the Cœlestiall Globe; and first of the Planets.

Stellis. HITHERTO hath our discourse beene concerning those things which are common to both Globes; we will now descend to speak of those that properly belong to each of them in particular. And first of those things that only concerne the Cœlestiall Globe; as namely the Stars, with their severall configurations.

The whole number of Starres hath been divided by the Ancient Astronomers, who first applied themselves to the diligent observing of the same, into two kinds. The first is of the Planets or wandring Starres; the other of the fixed. The first of which they therefore called Planets or Wanderers, because they observe no constant distance or situation, neither in respect of each other, nor in respect of those that are called fixed Starres. And these were so called because that they were observed alwayes to keep the same situation and distance from one another as is at large proved by Ptolomy in his Almagest, lib. 7, cap. 1, out of his owne observations, diligently compared with those delivered by Hipparchus.

Planetis. The Planets (excepting those two greater lights, the Sunne and Moone) are five in number. All which, beside the Diurnale motion, by which they are carried about from East to West by the Rapture of the first Movable, have also a free proper motion of their owne, which they finish from West to

East, according to the succession of the Signes upon the Poles of the Zodiaque, each of them in a severall manner and space of time; their order in the Heavens and period of their motions being such as followeth.

Saturne, called in Greeke *κρονος* or *φαινων* (and by ♄
Julius Higinus, Stella Solis, the Starre of the Sunne), is the highest of all the Planets, and goeth about the greatest circuit, but doth not therefore appeare to be the least of all the Planets, as Pliny thence conjectured. He finisheth his Periodicall course in twenty-nine yeares, five moneths, fifteen days, according to Alfraganus.

Jupiter, in Greek Σευς and *φαεθων*, moveth through the ♃
Zodiaque in the space of eleven years, tenne moneths, and almost 16 dayes.

Mars, 'Αρης and *πυτροεσις* (which is also called by some ♂
Hercules his Star), finishes his course in two yeares.

Sol, the Sunne, in Greek Ηλιος, performeth his course in ☉
a yeare, that is to say, three hundred sixtie five dayes and almost sixe houres.

Venus, Αφροδιτη (called by some Juno's Starre, by others ♀
Isis, and by others the Mother of the Gods), when it goeth before the Sunne it is called *φωσφορος*, the day Starre, appearing like another lesser Sunne, and as it were matural-ing the day. But when it followeth the Sunne in the Even-ing, protracting the light after the Sunne is set, and sup-plying the place of the Moone, it is then called Εσπερος, the Evening Starre. The nature of which Starre, Pythagoras Samius is said first to have observed about the thirtie 2d Olympiad, as Pliny relates, lib. 2, cap. 8. It performeth its course in a yeares space or thereabout, and is never distant from the Sunne above fortie sixe degrees, according to Timœus his computation. Notwithstanding our later Astro-nomers, herein much more liberall than hee, allow it two whole signes or 60 degrees, which is the utmost limit of its deviation from the Sunne.

☿ Mercury, in Greeke Ερμης and Στιλβων (called by some Apollos Starre), finisheth his course through the Zodiaque in a yeare also. And, according to the opinion of Timœus and Sosigenes, is never distant from the Sunne above 25 gr., or as our later writers will have it, not above a whole signe, or 30 degrees.

☽ Luna, Σεληνη, the Moone, is the lowest of all the Planets, and finisheth her course in twentie seven dayes and almost eight houres. The various shapes and appearances of which planet (seeming sometimes to bee horned, sometimes equally divided into two halves, sometimes figured like an imperfect circle, and sometimes in a perfect circular forme), together with the other diversities of this Starre, were first of all observed by Endymion, as it is related by Pliny; whence sprung that poetical fiction of his being in love with the Moone.

All the Planets are carried in Orbes which are Eccentrical to the Earth; that is, which have not the same center with the Earth. The Semidiameter of which Orbes, compared to the Semidiameter of the Earth, have this proportion as is here set downe in this table:

Of what parts the Semidiameter of the Earth is 1. Of the same the Semidiameter of the Orbe of—	Luna	is	48	56 m.
	Mercury		116	3 m.
	Venus		641	45 m.
	Sol		1165	23 m.
	Mars		5032	4 m.
	Jupiter		11611	31 m.
	Saturne		17225	16 m.

The Eccentricities of the Orbes compared with the Orbes themselves have this proportion.

Maurol. ex Alfonso.

Of what parts the Semidiameter of the Deferent is 60. Of the same the Eccentricity of—	Luna	is	12	28 m.	30 sec.
	Mercury		2	0 m.	
	Venus		1	8 m.	
	Sol		2	16 m.	6 sec.
	Mars		6	0 m.	
	Jupiter		2	45 m.	
	Saturne		3	25 m.	

The Eccentricities of some of the Planets (especially of

the Sunne) are found to have decreased and grown lesse since Ptolomies time. For Ptolomy sets downe the Eccentricity of the Moone to be 12 gr. 36 m., but by Alphonsus it was found to be but 13 gr. 28 m. and a halfe. Ptolomy assigned Eccentricity to Venus 1 gr. 14 m., Alphonsus 1 gr. 8 m. Ptolomy found by his owne observations, and also by those that Hipparchus had made, that the Eccentricity of the Sun was 2 gr. 30 m. Alphonsus observed it in his time **Fixis.** to be but 2 gr. 16 m. and 10th part of a minute. In the year of our Lord 1312, it was found to be 2 gr. 2 m. 18 sec. Copernicus found it to be lesse than that, and to be but 1 gr. 56 m. 11 sec. So that without just cause did the illustrious Julius Scaliger think Copernicus his writings to deserve the sponge, and the Author himselfe the bastinadò; herein dealing more hardly with Copernicus then he deserves.

CHAPTER II.

Of the Fixed Stars and their Constellations.

And here in the next place we intend to speake of the Fixed Stars, and their Asterismes or Constellations, which Pliny calls Signæ and Sidera Signes. Concerning the number of which Constellations, as also their figure, names, and number of the Stars they consist of, there is diversity of opinion among Authors. For Pliny, in his 2d book, 41 chap., reckoneth the whole number of the figure to be 72. But Ptolomy, Alfraganus, and those which follow them, acknowledge but 48 for the most part; notwithstanding some have added to this number one or two more, as Berenice's Haire, and Antinous. Germanicus Cæsar, and Festus Avienus Rufus, following Aratus, make the number lesse. Julius Higinus will have them to be but 42, reckoning the Serpent, and The Man that holdeth it for one Sign; and he omitteth the little Horse, and doth not number Libra among the Signes; but he divideth Scorpio into two Signes, as many others also doe. Neither doth hee reckon the Crow, the Wolfe, nor the South Crowne among his Constellations, but only names them by the way. The Bull also, which was described to appeare but halfe by Pliny and Hipparchus, and Ptolomy and those that follow them; the same is made to be wholly apparent both by Vitruvius and Pliny, and also before them by Nicander, if we may believe Theon, Aratus his Scholiast, who also place the Pleiades in his backe.

Concerning the number also of the Starres that goe to the making up of each Constellation, Authors doe very much differ from Ptolomy, as namely Julius Higinus, the Commentator upon Germanicus (whether it be Bassus, as Philander calls him, or whether those Commentaries were written by Germanicus himselfe, as some desire to prove out of Lactantius), and sometimes also Theon in his Commentaries upon Aratus, and Alfraganus very often.

Now, if you desire to know what other reason there is why these Constellations have beene called by these names, save onely that the position of the Starres doth in some sort seeme to expresse the formes of the things signified by the same; you may read Bassus and Julius Higinus, abundantly discoursing of this argument out of the fables of the Greekes. Pliny assures us (if at least we may believe him) that Hipparchus was the man that first delivered to posterity the names, magnitude, and places of the Starres. But they were called the same names before Hipparchus his time by Timochares, Aratus, and Eudoxus. Neither is Hipparchus ancienter than Aratus, as Theon would have him to be. For the one flourished about the 420 yeare from the beginning of the Olympiads, as appeareth plainely out of his life, written by a Greeke Author. But Hipparchus lived about 600 yeares after the beginning of the Olympiads, as his observations delivered unto us by Ptolomy doe sufficiently testifie. Besides that there are extant certaine Commentaries upon the Phenomena of Eudoxus and Aratus which goe under Hipparchus his name; unlesse perhaps they were written by Eratosthenes (as some rather thinke), who yet was before Hipparchus.[1]

Pliny, in his 2 booke, 41 chapter, affirmeth (though I know not upon whose authority or credit) that there are reckoned 1600 fixed Starres, which are of notable effect and vertue. Whereas Ptolomy reckoneth but 1022 in all, accounting in those which they call Sporades, being scattered here and there and reduced to no Asterisme. All which, according to their degrees of light, he hath divided into 6 orders. So that of the first Magnitude he reckoneth 15; of the second, 45; of the third, 208; of the fourth, 474; of the fifth, 217; of the sixth, 49; to which we must add the 9 obscure ones, and 5 other which the Latines called Nebulosæ, cloudy Starres. All which Starres expressed in their severall Constellations, Magnitudes, and Names, both in Latine and Greeke (and some also with the names by which they are called in Arabique), you may see described in the Globe.

All these Constellations (together with their names in Arabique, as we find them partly set downe by Alfraganus, partly by Scaliger in his Commentaries upon Manilius, and Grotius his notes upon Aratus his Asterismes, but especially Jacobus Christmannus hath delivered them unto us out of the Arabique epitome of the Almagest) we will set downe in their order. And if any desire a more copious declaration of the same, we must refer him to the 7 and 8 booke of Ptolomies Almagest, and Copernicus his Revolutions, and the Prutenicke Tables digested by Erasmus Reinholt; where every one of these Starres is reckoned up, with his due longitude, latitude, and magnitude annexed.[2]

[1] Pontanus refers to the conjecture that the stars were reduced into constellations by two kinds of men, husbandmen and mariners: and to the names of stars in the translations of Job.

[2] Pontanus also refers the reader to the commentary on Sacrobosco by Clavius, and above all to Tycho Brahe.

But here you are to observe by the way Copernicus and Erasmus Reinholt doe reckon the longitude of all the Starres from the first star in Aries; but Ptolomy from the very intersection of the Æquinoctiall and Eclipticke. So that Victorinus Strigelius was in an error when he said that Ptolomy also did number the longitude of Starres from the first Starre, the head of Aries.

Strig. de primo motu parte tertia.

CHAPTER III.

Of the Constellations of the Northerne Hemisphere.

Asterismi enumerati.

The first is called in Latine Ursa Minor, and in Arabique Dub Alasgar, that is to say, the lesser Beare, and Alrucaba, which signifieth a Wagon or Chariot; yet this name is given also to the hinder most Starre in the taile which in our time is called the Pole Starre, because it is the nearest to the Pole of any other. Those other two in the taile are called by the Greekes *χορευται*, that is to say, Saltatores, Dancers. The two bright Starres in the fore part of the body the Arabians call Alferkathan, as Alfraganus writeth, who also reckoneth up seven Starres in this Constellation, and one unformed neare unto it. This Constellation is said to have been first invented by Thales, who called it the Dog, as Theon upon Aratus affirmeth.

The second is Ursa Major, the Great Beare; in Arabic, Dub Alacber. The first Starre in the backe of it, which is the 16 in number, is called Dub, *κατεξοχιω*, and that which is in the flanke, 17 in number, is called Mirae, or rather, as Scaliger would have it, Mizar, which signifieth (saith he) *locum præcinctionis*, the girthing place. The first in the taile, which is the 25 in number, is called by the Alfonsines Aliare, and by Scaliger Aliath. This Asterisme is said to have beene first invented by Naplius, as Theon affirmeth.

It hath in all 27 Starres, but as Theon reckoneth them, but 24. Both the Beares are called by the Greekes, according to Aratus, *ἁμαξα*, which signifieth a Wagon or Chariot. But this name doth properly appertaine to those seven bright Starres in the Great Beare which doe something resemble the forme of a wagon. These are called by the Arabians Beneth-As, *i.e.*, Filiæ Feretri, as Christmannus testifieth. They are called by some, though corruptly, Benenas, and placed at the end of the taile. Some will rather read it Benethasch, which signifies Filiæ Ursæ. The Grecians in their Navigations were wont alwayes to observe the Great Beare, whence Homer gives them the Epithete *ελικωπας* as Theon observeth, for the Greekes call the Great Bear *ελικη*. But the Phœnicians alwayes observed the lesser Beare, as Aratus affirmeth.

The third is called the Dragon, in Arabique Alanin, and it is often called Aben; but Scaliger readeth it Taben; whence hee called that Starre which is in the Dragons head, and is 5 in number, Rastaben, though it be vulgarly written Rasaben. In this Constellation there are reckoned 31 Starres.

The fourth is Cepheus, in Arabique Alredaf. To this Constellation, besides those two unformed Starres which are hard by his Tiara, they reckon in all 11, among which that which is in number the 4 is called in Arabique Alderaimin, which signifieth the right Arme. This Constellation is called by the Phœnicians Phicares, which is interpreted Flammiger, which appellation peradventure they have borrowed from the Greeke word *πυρκαεις*.

The fifth is Bootes, *βοωτης*, which signifieth in Greeke an Heardsman, or one that driveth Oxen. But the Arabians mistaking the word, as if it had been written *βοατης* of *βοαω*, which signifies *Clamator*, a Cryer, call it also Alhava, that is to say, Vociferator, one that maketh a great Noyse or Clamor; and Alsamech Alramech, that is, the

Launce bearer. Betwixt the legs of this Constellation there stands an unformed star of the first magnitude, which is called both in Grecke and Latine Arcturus and in Arabique Alramech, or the brightest Starre, Samech haramach. This Starre Theon placeth in the midst of Bootes his belt or girdle. The whole Constellation consisteth of 22 Starres.[1]

The sixth Constellation is Corona Borea, the North Crowne, called by the Arabians Aclilaschemali, and that bright Starre which is placed where it seemeth to be fastened together, and which is the first in number, is called in Arabique Alphecca, which signifieth *Solutio,* an untying or unloosing. It is also called Munic; but this name is common to all bright Starres. The whole Constellation consisteth of eight Starres.

The seventh is Hercules, in Arabique Alcheti hale rechabatch, that is, one falling upon knees, and sometimes absolutely Alcheti, for it resembles one that is weary with labour (as Aratus conceives), whence it is also called in Latine Nisus or Nixus (which in Vitruvius is corrupted into Nesses), and the Greeks call it *ενγονασι*, that is to say, One on his knees. The Starre which is first in number in the head of this Constellation is called in Arabique Rasacheti, not Rasaben, as the Alfonsines corruptly have it; and the 4 Starre is called Marsic, or Marfic *Reclinatorium,* that part of the Arme on which we leane. The eight Starre, which is the last of the three, in his Arme, is called Mazim, or Maasim, which signifieth Strength. This Constellation hath eight Starres, besides that which is in the end of his right foote, which is betwixt him and Bootes, and one unformed Starre at his right Arme.

The eight is the Harpe, called in Latine Lyra, in Arabique Schaliaf and Alvakah, *i.e.*, Cadens, *sc.* Vultur, the

[1] **Pontanus discusses the word Arcturus, and mentions that the word in Job, which is given as Arcturus in the Septuagint, is *Ash* in Hebrew, from the root *Gnusch* ("*congregabit*").**

falling Vulture. It consisteth of ten Starres, according to Hipparchus and Ptolomy; but Timochares attributed to it but 8, as Theon affirmeth, and Alfraganus 11. The bright, Starre in this Constellation, being the first in number, Alfonsus calleth Vega.

The ninth is Gallina or Cygnus, the Hen or Swan, and is called in Arabique Aldigaga and Altayr, that is, the flying Vulture. To this Asterisme they attribute, besides those two unformed neare the left wing, 17 Starres, the 5 of which is called in Arabique Deneb Adigege, the taile of the hen, and by a peculiar name Arided, which they interpret *quasi redolens lilium*, smelling as it were of lilies.[1]

The 10th is Cassiopeia, in Arabique Dhath Alcursi, the Ladye in the Chayre; and it consisteth of 13 Starres, among which the 2d in number Alfonsus calleth Scheder, Scaliger Seder, which signifieth a breast.[2]

The 11th is Perseus, Chamil Ras Algol, that is to say, bearing the head of Medusa; for that Starre which is on the top of his left hand is called in Arabic Ras Algol, and in Hebrew Rosch hasaitan, the Divels Head. This Constellation hath, besides those three unformed, 26 other Starres; of which that which is the seventh in number Alfonsus calleth Alchcemb for Alchenib, or Algeneb, according to Scaliger, which signifieth a side.

The 12th is Auriga the Wagoner, in Arabique Roha, and Memassich Alhanam. That is one holding the raines of a bridle in his hand. This Asterisme hath 14 Stars; of which that bright one in the left shoulder, which is also the third in number, is called in Greeke αιξ, Capra, a Goate; and in Arabique Alhaisk, or, as Scaliger saith, Alatod, which signi-

[1] Pontanus here mentions the appearance of a new star in the breast of the swan, in 1600, which was observed by Kepler and others.

[2] A new star which appeared in Cassiopeia, in 1572, is here referred to by Pontanus.

fieth a He Goate; and the two which are in his left hand, and are 8th and 9th, are called εριφοι, Hœdi, Kids; and in Arabique, as Alfonsus hath it, Saclateni; but according to Scaliger, Sadateni, the hindmost arme. This Configuration of these Starres was first observed by Cleostratus Tenedius, as Higinus reporteth.

The 13th is Aquila, Alhakkah, the Eagle; the moderne Astronomers call it the flying Vulture, in Arabique Altayr; but Alfraganus is of a contrary opinion, for he calleth the Swanne by this name, as we have already said. They reckon in this Asterisme 9 Starres, besides 6 unformed,
Antinous. which the Emperor Hadrian caused to be called Antinous, in memory of Antinous his minion.

The 14th is the Dolphin, in Arabique Aldelphin, and it hath in it 10 Stars.

The 15th is called in Latine Sagitta or Telum, the Arrow or Dart, in Arabic Alsoham; it is also called Istuse, which word Grotius thinkes is derived from the Greeke word οισος, signifying an arrow. It containeth 5 Stars in all.

The 16th is Serpentarius, the Serpent bearer, in Arabic Alhava and Hasalangue. It consisteth of 24 Starres, and 5 other unformed. The first Starre of these is called in Arabique Rasalangue.[1]

The 17th is Serpens, the Serpent, in Arabique Alhasa; it consisteth of 18 Starres.

The 18th is Equiculus, the little Horse, and in Arabique Katarat Alfaras, that is in Greeke προταμη ιππος, as it were the fore part of a Horse cut off. It consisteth of 4 obscure Starres.

The 19th is Pegasus, the Great Horse, in Arabique Alfaras Alathem; and it hath in it 10 Stars. The Starre on the right shoulder, which is called Almenkeh, and is the third in number, is also called Seat Alfaras, Brachium Equi.

[1] In 1605 a new star was discovered in the foot of Serpentarius, which disappeared in 1606. Kepler wrote a treatise on it.

And that which is in the opening of his mouth, and is numbered the 17th, is called in Arabique Enif Alfaras, the nose of the Horse.

The 20th is Andromeda, in Arabique Almara Almasulsela, that is, the Chained Woman; Alfraganus interprets it Fæminam quæ non est experta virum: A Woman that hath not knowen a man. This Constellation containeth in it 23 Stars; whereof that which is the 12th in number, and is in the girdling place, is commonly called in Arabique Mirach, or, according to Scaliger, Moza; and that which is the fifth is called Alamec, or rather Almaac, which signifies a socke or buskin.

The 21st is the Triangle, in Arabique Almutaleh and Mutlathun, which signifies Triplicity. It consisteth of 4 Starres.[1]

CHAPTER IV.

Of the Northerne Signes of the Zodiaque.

The first is Aries, the Ram, in Arabique Alhamel; this Constellation hath 13 Starres, according to Ptolomies account. Yet Alfraganus reckoneth but 12, beside the other 5 unformed ones that belong to it.

The 2d is Taurus, the Bull, in Arabique Altor or Ataur; in the eye of this Constellation there is a very bright Star, called by the Ancient Romans Palilicium, and by the Arabians Aldebaram, which is to say, a very bright Star, and also Hain Altor, that is, the Bull's Eye. And those five Stars that are in his forehead, and are called in Latine Suculæ, the Grecians call *υαδες*, because, as Theon and Hero Theon in Aratum.

[1] Pontanus says that the whole number of stars in the northern part of the heaven is 360, of which only three are of the first magnitude, Capella, Vega, and Arcturus.

Mechanicus conceive, they represent the forme of the letter ♈; although perhaps it is rather because they usually cause raine and stormy weather. Thales Milesius said that there were two of these Hyades, one in the Northerne Hemisphere and one in the South; Euripides will have them to be 3, Achæus 4, Hippias and Pherecides 7. Those other 6, or rather 7 Stars that appeare on the back of the Bull, the Greekes call Pleiades (perhaps from their multitude); the Latines Vergiliæ; the Arabians Atauriæ, quasi Taurinæ, belonging to the Bull. Nicander, and after him Vitruvius, and Pliny place these Stars in the taile of the Bull; and Hipparchus quite out of the Bull, in the left foot of Perseus. These Stars are reported by Pliny and Solinus to be never seene at all in the Isle Taprobana; but this is ridiculous, and fit to bee reported by none but such as Pliny and Solinus. For those that inhabite that Isle have them almost over their heads. This Constellation hath 33 Stars in it, besides the unformed Stars belonging to it, which are 11 in number.[1]

The third is Gemini, the Twinnes, in Arabique Algeuze. These some will have to bee Castor and Pollux, and others Apollo and Hercules; whence, with the Arabians, the one is called Apellor or Apheleon, and the other Abracaleus, for Gracleus, as Scaliger conceiveth. It containeth in it (beside the 7 unformed) 18 Stars, amongst which that which is in their head is called in Arabique Rasalgeuze.

The fourth is Cancer, the Crab, in Arabique Alsartan; consisting of 9 Stars, beside 4 unformed; of which that cloudy one which is in the breast, and is the first of all, is called Mellef in Arabique, which, as Scaliger saith, signifieth thicke or well compact.

The fifth is Leo, the lion, in Arabique Alased, in the breast whereof there is a very bright Starre, being the 8th in number, and is called in Arabique Kale Alased, the

[1] Pontanus says that the words of Pliny do not convey the sense attributed to them in the text

heart of the Lion, in Greeke βασιλιοκος, because those that are borne under this Starre have a Kingly Nativity, saith Proclus. And that which is in the end of the taile, and is the last of all in number, is named Deneb Alased, that is, the taile of the Lion; Alfraganus calleth it Asumpha. This Constellation containeth in it 27 Stars, besides 8 unformed. Of the unformed Stars, which are betwixt the hinder parts of the Lion and the Great Beare (according to Ptolomies account, although Theon, following Aratus, reckons the same as belonging to Virgo), they have made a new Constellation, which Conon the Mathematician, in favour of Ptolomy and Berenice, would have to bee called Berenice's Haire; which story is also celebrated by the Poet Callimachus in his verses.

Proclus de Sphæra.

The sixth is Virgo, the Virgin, in Arabique Eladari; but it is more frequently called Sunbale, which signifieth an Eare of Corne; and that bright Starre which she hath in her left hand is called in Greeke σαχυς, an Eare of Corne, and in Arabique Hazimeth Alhacel, which signifieth an handfull of Corne. This Star is wrongly placed by Vitruvius and Higinus in her right hand. The whole Constellation consisteth of 26 Stars, besides the 6 unformed.

CHAPTER V.

Of the Constellations of the Southerne Hemisphere: and first of those in the Zodiaque.

And first of Libra, which is the 7 in order of the Signes. That part of this Constellation which is called the Southerne Ballance, the Arabians call Mizan Aliemin, that is to say, Libra dextra vel meridionalis, the Right hand or Southerne Ballance. But Libra was not reckoned anciently among the

Signes; till that the later Astronomers, robbing the Scorpion of his Clawes, translated the same to Libra, and made up the number of the Signes, whence the Arabians call the Northerne Ballance Zubeneschi Mali, that is in Greeke, *χηλη βορειȣ*, the North Clawe; and the other part of it that looks Southward they call Zubenalgenubi, *χηλη νοτιου*, the South Claw. This Constellation containeth in it 8 Starres, besides 9 other unformed, belonging unto it.

The Eight is Scorpio, the Scorpion, in Arabique commonly called Alatrah, but more rightly Alacrah; whence the Starre in the breast of it, which is the 8 in number, is called Kelebalacrah, that is, the Heart of the Scorpion; and that in the end of his taile, which is the second in number, they call Leschat, but more truly Lesath, which signifieth the sting of any venomous creature; and by this word they understand the Scorpions sting. It is also called Schomlek, which Scaliger thinks is read by transposition of the letters for Moselek, which signifieth the bending of the taile. This Constellation consisteth of 21 Starres, besides 3 unformed.

The ninth is Sagittarius, the Archer, in Arabique Elcusu or Elcausu, which signifieth a Bow; it hath in it 31 Starres.

The tenth is Capricornus, the Goat, in Arabique Algedi. To this Constellation they attribute 28 Starres, among which that which is in number the 23 is called in Arabique Deneb Algedi, the taile of the Goat.

The eleventh is Aquarius, the Waterman, in Arabique Eldelis, which signifieth a bucket to draw water. The 10 Starre of this Constellation is called in Arabique Seat, which signifieth an Arme. It containeth in all 42 Stars.

The Twelfth is Pisces, the Fishes, in Arabique Alsemcha. It containeth 34 Starres, and 4 unformed.[1]

[1] Pontanus reckons the number of zodiacal stars at 346, of which only five are of the first magnitude—Aldebaran, Regulus, Cauda Leonis, Spica, and a star near the mouth of the southern fish.

CHAPTER VI.

Of the Constellations of the Southerne Hemisphære, which are without the Zodiaque.

The first is Cetus, the Whale, called in Arabique Elkaitos, consisting of 22 Starres. That which is in number the second is commonly called Menkar, but more rightly, as Scaliger saith, Monkar Elkaitos, the nose or snout of the Whale; and the 14, Boten Elkaitos, the belly of the Whale; and the last of all save one, Deneb Elkaitos, the taile of the Whale.

The second is Orion, which the Arabians call sometimes Asugia, the Mad Man; which name is also applied to Hydra, and sometimes to Elgeuze. Now, Geuze signifieth a walnut, and perhaps they allude herein to the Latine word Ingula, by which name Festus calleth Orion; because he is greater then any other of the Constellations, as a walnut is bigger than any other kinde of nut. The name Elgeuze is also given to Gemini. This Constellation is also called in Arabique Algibbar, which signifies a strong man or Gyant. It consisteth of 38 Starres, among which that which is the second, and is placed in his right shoulder, is called Jed Algeuze, that is, Orion's Hand, as Christmannus thinketh: but more commonly Bed Elgeuze, and perhaps it should rather be Ben Elgeuze, that is, the bright Starre in Orion. The third Starre is called by the Alfonsines Bellatrix, the Warrior. That which is in his left foote, and is the 35 in number, Rigel Algeuze or Algibbar, that is to say, Orion's foote.[1]

The third is Eridanus, in Arabique Alvahar, that is to say, the River; whence Nar, the name of a River in Hetruria, is conceived by some to have been contracted. It hath in it

[1] Pontanus here again alludes to the mention of Orion in the translations of Job. The Hebrew word is *Kesil*, which means rage or madness, answering to the Arabic *Asugia*.

34 Starres; among which that which is the 19 is commonly called in Arabique Angetenar, but Scaliger rather thinks it should be read Anchenetenar, which signifieth the winding or crooking of a River. The 29 Starre is also called Beemim, or rather Theemim, which signifieth any two things joyned together, so that it is to be doubted whether or no this name may not be as well applied to any two Starres standing close by one another. And the last bright Starre in the end of it is called Acharnahar, as if you should say Behinde the River, or in the end of the River, and it is commonly called Acarnar.

The fourth is Lepus, the Hare, in Arabique Alarnebet and it containeth in all 22 Stars.

The fifth is Canis, the Dogge; Alcheleb, Alachbar, in Arabique, the great Dog; and Alsahare aliemalija, that is to say, the Right hand or Southerne Dog. Which name Alsahare, which is also sometime written Scera, Scaliger thinkes is derived from an Arabique word which signifieth the same that *υδροφοβια* in Greeke, a disease that mad dogs are troubled with, when as they cannot endure to come neare any water. Notwithstanding, Grotius is in doubt whether or no it should not rather be Elseiri, and so derived from the Greeke word *σειριος*. For by this name is that notable bright Starre called which is in the Dogs mouth, and is called in Arabique Gibbar or Ecber, and by corruption Habor. This Constellation hath in it 11 Stars.

The sixth is the little Dog, called in Greeke Procyon, and in Latine Antecanis, because it riseth before the great Dog. The Arabians call it Alcheleb Alasgar, that is to say, the lesser Dog, and Alsahare Alsemalija, and commonly though corruptly Algomeiza, the left hand or Northerne Dog. This Asterisme consisteth of two Stars onely.

The seventh is Argo, the Shippe, in Arabique Alsephina; now Sephina signifieth a Ship. It is also called Merkeb, which signifieth a Chariot; according as the Poets also usually cal it *αρμα θαλασσης*, as if one should say a Sea chariot instead of a Ship. But the Alphonsines give this appellation to that Star which is the 6 in number. The whole Asterisme containeth in it 45 Stars, of all which that which is the last save one is called in Arabique Sohel or Syhel, which signifieth ponderous or weighty, which appellation they perhaps have given it for the same reason that Bassus hath another like it, which is Terrestris, because it alwayes appeareth to them very low, and neare the earth. The Greeks call this Star *κανωβος*, the Hebrewes Chesil, as Christmannus is of opinion. Which, if it be so, then Arias Montanus is in an errour in taking it for Orion in his translation of the Itinerary of Beniamin Tudelensis. The inhabitants of Azania called it an Horse, as Ptolomy affirmes in his Geogr., lib. 5, cap. 7. C. 7. l. 5. Geograph.

The eight is Hydra, in Arabique Alsugahh or Asuta, which signifieth strong or furious. The Egyptians call it Nilus, as Theon writeth in his Commentaries upon Aratus. It hath in it 25 Starres, besides two unformed; the 12 of which the Alfonsines call Alphart.

The ninth is Crater, the Cup, in Arabique Albatina and Elkis, which signifieth a Goblet or standing Cup. It hath in it 7 Stars.

The tenth is Corvus, the Crow; Algorab in Arabique, consisting of 7 Starres.

The eleventh is Centaurus, the Centaur; called by the same name in Arabique. It containeth 37 Stars; among which those that are in his hinder feete are the Stars that make up the Crosse, so much celebrated in the Spanish Navigations.

The twelfth is Fera, the Wild beast, called in Arabique Asida, signifying a Lionesse; and Alsubahh, which also is taken for a Wolfe or other ravenous beast. To this Constellation they reckon 19 Stars.

The Thirteenth is Ara or Thuribulum, the Altar or Censer.

in Arabique Almugamra; Bassus calls it Sacrarium. It containeth 7 Stars.

The foureteenth is Corona Australis, or South Crowne, in Arabique Alachil Algenubi. It consisteth of 13 Stars, making up a double wreath, according to Alfraganus; yet Theon reckoneth but 12 in it.

The fifteenth is Piscis Austrinus, the South Fish; Ahaut Algenubi, in Arabique. It containeth in it 12 Starres in Ptolomies account, but 11 onely according to Alfraganus. Among which the bright one that is in his mouth is called Phom Ahut, that is to say, the mouth of the Fish; and commonly by corruption Fomahaut. There is also described in the Cœlestiall Globe a certaine broad Zone or circle of the colour of milke, which representeth that which appeareth in the Heavens, and is commonly called Via Lactea, the Milky Way. Which Zone or circle is not drawne regularly or equally either in respect of latitude, colour, or frequency of Stars; but is different and various both in forme and situation, in some places appearing but as a single circle, and againe in others seeming as it were dividing in two parts. The delineation whereof you may see in the Globe, and the description more largely set down by Ptolomy in his Almagest, lib. 8, cap. 2.[1]

CHAPTER VII.

Of the Starres which are not expressed in the Globe.

Besides those Starres which we have here reckoned up out of Ptolomy, there are yet many other to bee seene sometime, especially in the winter time in a cleare night, when as

[1] Pontanus gives 316 as the number of stars in the southern heaven, those of the first magnitude being Betelguese, Rigel, Achenar, Sirius, Procyon, Canopus, and a star in the right foot of the Centaur.

there are both many more Stars to be seene then at any other times, and those that are seene appeare by much greater. Now, if you expect that we should assigne the cause of this, we might answer that it is beside the intention of our present purpose. Yet for your satisfaction, and because that some authors have very much erred from the right in setting downe the true reason of the same, we doe therefore the more willingly make this digression. For some there are who (out of the extraordinary knowledge they have in Philosophy and Optickes) would very willingly perswade us that either we conceive them to be more then indeed they are, and that our sense onely is deceived, or else (which is altogether as ridiculous) that the ayre being in winter more pure and thin, maketh them more conspicuous, which otherwise in the summer, when the ayre is more grosse, doe altogether lye hid. And this is an error which I doe not so much blame in others, as I wonder at it in Johannes Benedictis, that so great a Mathematician as he is held to be should be led away with so grosse an error. For the reason of this is altogether otherwise and cleane contrary. For that very cause that the ayre is more grosse and thicke, the Stars therefore doe appeare more and greater. Which opinion of ours is confirmed, both out of principles of the Optickes, and also by the sense of it selfe, experience, and authority of learned writers.

For first, that the raies being refracted through a grosse Medium, and diffused as it were into certaine Canales, doe represent the image of the object greater then indeed it is, is plainely affirmed (and that according to the doctrine of the Optickes) by Strabo himselfe out of Posidonius. And that L. 3.
through Perspicills or Spectacles things appear more and greater then otherwise they would, is a thing well known to the most Ignorant. Cleomedes also saith, that the Sunne Cleomedes, lib. 1.
being seene by any in the bottome of a deepe well seemes greater then when he is seene from above: and that by

reason of the moystnes and grossenesse of the ayre in the bottome of the Well. And if it were possible to see the Sunne through stone walles or other solid bodies (as the old Poets fabulously report of Lynceus), he would seeme much bigger then he is, as Posidonius rightly teacheth. And hence is it, saith Strabo, that we see the Sunne alwayes greater at his rising and setting, especially to those that are at Sea. Yet we doe not say that he appeares ten times greater then he is, as it is reported he doth in India, out of the excerpts of Etesias his Indian Histories; much lesse that he seemes to be an hundred times greater then he is in other places, as he is feigned by Artemidorus to be at his setting, to those that inhabit a Promontory in the outmost C. 5, Vincens. parts of Spaine, which he calls Promontorium Sacrum; but Alfrag., c. 2. is justly taxed for the same by Posidonius. Alfraganus would have the cause of this to be, for that the vapours which are exhaled out of the earth, and elevated into the ayre, and so interposed betwixt our sight and the Sunne at his rising or setting, doe make him appear greater then he really is. The same is the opinion of Strabo and Cleomedes, also out of Posidonius; neither doth this differ much from the opinion of the best of our Opticall writers. But of this enough.

There are also observed many Stars in the Southerne Varie Relationes stel. Aust. parts of the World, which, because they could not be seene by our Artists in this part of the world, we have therefore no certaine knowledge left us concerning the same. So in like manner among those which we have hitherto spoken of, many of them cannot be seene by those that inhabite any whit nearer the North Pole. But concerning those Stars that appeare about the South Pole of the world, I will here set you downe a very admirable story which Franciscus Patricius. Patricius Senensis relateth in the end of his *Nova Philosophia*, out of the Navigationes of Americus Vespuccius. And it is thus: Cœlum decentissime exornatur, etc. The Heavens (saith he, meaning about the Antarctique Pole) is variously adorned with diverse Constellations which cannot be seene here with us; among which I doe very well remember that I reckoned very neare twenty which were as faire and bright as Venus and Jupiter here with us. And a little after he saith: I was certaine, therefore, that these Stars were of greater Magnitude then any man can conceive; and especially three Canobi, which I saw and observed; two whereof were very bright ones, but the third was somewhat obscured, and nothing like the rest.

And a little after he proceeds: But the Pole it selfe is encompassed about with three Stars, which represent the figure of a right angled Triangle; among which that which is in the midst is in circumference 9 gr. and a halfe; and when these rise there appeares on the left hand of them another bright Canobus of notable magnitude.

And a little after he saith: After these there follow three other very faire Stars, the middlemost of which hath in Diameter 12 degrees and an halfe; and in the midst among these is seen another Canobus. After this there follow 6 other bright Stars which excel all the other Stars in the eighth Sphære for brightnesse; the middlemost of them having 32 gr. in Diameter. These Stars were accompanied by another great but darker Canobus; all which Stars are observed in the Milky Way.

To this he addeth out of Corsalius that which followeth: Andreas Corsalius also affirmeth that there are two clouds, of a reasonable brightnesse, appeareing near the Pole; betwixt which there is a Star, distant from the Pole about 11 gr., over which he saith there is seene a very admirable figure of a Crosse standing in the midst of 5 Stars that compasse it about, with some certaine others that move round about with it, being distant from the Pole about 30 degrees; which are of so great brightnesse as that no Signe in the Heavens may be compared with them.

And now that you have heard this so strange and admirable relation of the Stars about the Antarctique Pole, Auditum admissi risum teneatis? For Vespuccius here hath forged three Canobi, whereas Ptolomy and all the Ancient Greekes never knew but one, and that is it which is placed in the sterne of the ship Argo. And here it is very well worth our noting, that Patricius (as farre as I am able to gather out of his writings), out of Vespuccius his ill-expressed language, and by him worse understood, hath very excellently framed to himselfe a strange kinde of Star that hath in apparent Diameter 32 degrees; whereas the Diameter of the Sunne itselfe hardly attaineth to 32 minutes.

But those things which out of our owne certaine knowledge and experience in above a yeares voyage in the yeares 1591 and 1592, we have observed beyond the Æquator and about the Southerne parts of the world, we will here set downe.

Now, therefore, there are but three Stars of the first magnitude that I could perceive in all those parts which are never seene here in England. The first of these is that bright Star in the sterne of Argo which they call Canobus. The second is in the end of Eridanus. The third is in the right foote of the Centaure. To which if you will add for a fourth that which is fixed on the Centaures left knee, I shall not much stand against it. But other stars of the first magnitude then those which I have named that part of the world cannot shew us. Neither is there to be found scarcely two or three at the most of the second magnitude but what Ptolomy had seene. And, indeed, there is no part of the whole Heavens that hath so few Stars in it, and those of so small light, as this near about the Antarctique Pole. We had a sight also of those clouds Andreas Corsalius speakes of, one of them being almost twice or thrice as big as the other, and in colour something like the Via Lactea, and neither of them very far distant from the Pole. Our mariners used to call them Magellanes Clouds. And we saw also that strange and admirable Crosse that he talkes of, which the Spaniards call Crusero and our Countrimen the Crusiers. And the Stars of which this Crosse consists were not unknowne to Ptolomy also; for they are no other then the brighter Stars which are in the Centaures feete. And which thing I did the more diligently and oftener observe, for that I remembered that I had read in Cardan also strange relations of the wonderfull magnitude of the Stars about the South Pole, not unlike the stories we have now alleadged out of Patricius.

Card. de subtil.

THE THIRD PART.

CHAPTER I.

Of the Geographicall description of the Terrestriall Globe; and the parts of the world yet knowne.

Geographia Globi Terrestris.

DIONYSIUS AFER, in the beginning of his Periegesis, saith that the whole Earth may be said to be as it were a certaine vast island encompassed about on every side by the Ocean. The same was the opinion of Homer also before him, and of Eratosthenes (whom Dionysius is observed by Eustathius his Scholiast to follow in many things), as is witnessed by Strabo. The same is affirmed by Mela also after him. This vast Hand of the whole earth they would have to be terminated on the North side with the frozen Sea, which is called by Dionysius Mare Saturninum, and Mortuum; on the East with the Easterne Sea, which is also called Mare Sericum; on the South with the Red Sea (which Ptolomy calleth the Indian Sea) and the Æthiopian; and on the West with the Atlanticke Ocean. But of this Ocean also there are foure principall Gulfes (as the Ancient Geographers conceived) which embosomed themselves into the Maine land. Two of which derived their course out of the Erythræan or Red Sea, to wit, the Persian and Arabian Gulfes. From the West there is sent out of the Atlanticke Ocean a vast gulfe, which is called the Mediterranean Sea. And out of the North they would have the Scythian Ocean to send in the Caspian Sea, which is shut in almost on every side with high craggy rockes, from whence the streames flow with such violence that when they are come to the very fall they cast forth their water so farre into the Sea, without so much as once touching upon the Shore, that the ground is left dry and passeable for whole Armies under the bankes; the streames in the meane time being carried over their heads, as it is reported by Eudoxus in Strabo. This Sea, both Strabo, Pliny, Mela, and Solinus will have to come out of the Scythian Ocean (as we have said). But this errour of theirs, besides the experience of these later times, is manifestly convinced by this one testimony of Antiquity, which is that the water of this Sea is found to bee fresh and sweet, as was first observed by Alexander the Great and afterwards by Pompey, as M. Varro in Solinus testifieth, who at that time served under Pompey in his Warres. And this is the chiefest reason which Polycletus in Strabo alleadged for the proofe of the same.

Now all this tract of land the Ancients divided into two parts onely, namely, Asia and Europe, to which succeeding times added a third, which they called Africa, and sometimes also Libya. And of these Asia is the greatest, Africa the next, but Europe least of all; according as Ptolomy determines it in the 7 booke of his Geography.

Europe is divided on the East from Asia by the Ægean Sea (which is now called the Archipelago) and the Euxine Sea, which was at first (as Strato in Strabo thought) encompassed about on all sides in manner of a great lake, till at last by the great accession of other Rivers and waters it so far encreased as that the bankes being unable to containe it, it violently made its way into the Propontis and the Hellespont. The Euxine Sea is now called Mare Maggiore. It is also bound on the same side by the lake of Mœotis (now called Mare delle Zabacche), the river Tanais, now called Don, and the Meridian, which extends it selfe from thence to the Scythian or Frozen Sea. On all other sides it is encompassed with the Sea. For toward the South it is divided from Africa by the Straits of Gibraltar and part of the Mediterranean Sea.

Europa.

The length of these Straits is, according to Strabo and Pliny, 120 furlongs, and the breadth of it, according to the same Strabo, 70 furlongs. But Mela would have it to be 10 miles, that is to say, 80 furlongs. T. Livius and Cornelius Nepos make the latitude of it to be in the broadest place 10 miles or 80 furlongs; and where it is narrowest, 7 miles or 56 furlongs. But Turannius Graccula, who, as Pliny reports, was born about those parts, accounted it to be from Mellaria, a towne in Spaine, unto that promontory in Africa, which is called Promontorium Album, but 5 miles in all, that is, 40 furlongs. Eratosthenes was of opinion that Europe was sometime joyned to the Continent of Africa. And it is reported by Pliny that the inhabitants of those parts have a tradition that the Isthmus, or necke of the lande by which Europe and Asia were joyned together, was cut through by Hercules.

Europe is terminated on the West with the Atlanticke Ocean; and on the North with the British, Germane, and frozen Seas.

Africa.

Africa is divided from Asia (according to Dionysius and Mela) by the River Nilus, and a Meridian drawne through it to the Æthiopian Ocean. But Ptolomy would rather have its limits on this part to be the Arabian gulfe (which he not so rightly called the Red Sea), and a Meridian which should be drawne from thence to the Mediterranean Sea, over that necke of land which lyeth betwixt the two Seas, and which joyneth Ægypt to the Continent of Arabia and Indæa. Neither doth he thinke it congruous that Ægypt should be divided into two parts, one whereof should be reckoned to Africke, and the other to Asia; which must needs be if the river Nilus be set for the bounds of the same. Neither doth Strabo conceive this to be any whit improper, since that the length of the Isthmus which divideth the two Seas is not above a 1000 furlongs. And he seemeth to have said very rightly that it is not above a 1000 furlongs. For however

Posidonius reckoneth it to be very neere 1500 furlongs; yet Pliny would have it to be no more than 115 miles, that is to say, 920 furlongs. And Strabo also reckoneth the distance betwixt Pelusium and the Heroes city, which is situate close by the highest part of the Arabian gulfe, to be but 900 furlongs. But if we will give any credit to Plutarch, at the narrowest part of the Isthmus the two Seas will be found distant not above 300 furlongs. And that (when Anthony was overthrowne by Augustus in a Sea fight, and all his forces cleane broken) Cleopatra, seeking to avoid the servitude of the Romans, went about to transport her Navie this way over the firme land, that so she might finde some new place of habitation as farre remote from the Romans as she might; as it is reported by the same author, in the life of Anthony. But what should move Copernicus, in his 1 booke, 3 chap., to say that these two seas are scarcely 15 furlongs distant, I cannot conjecture; unlesse I should thinke the place to be corrupted through the negligence of the Transcribers or Printers. And yet I could wish that this (though it be a very great one) were all the errours that were to bee found in the writings of that most excellent man.

This Isthmus, as Eratosthenes conceived, was anciently covered all over with waters, till such time as the Atlanticke Ocean had intercourse with the Mediterranean. And some of the old Grammarians, Scholiasts on Homer, doe affirme (as Strabo testifieth) that it was this way that Menelaus in Homer sailed to the Æthyopians. I will therefore here set downe some few things which may seeme to make for the confirmation of this relation (whether you will call it an History or rather a Fable, or Conjecture) of Eratosthenes.

First, therefore, that Egypt (if not all of it, yet at least that part of it which is situated beneath Delta, and is called Egyptus Inferior, the lower Egypt, and is accounted to be the Gift of Nilus, or rather of the Sea) was made by the aggestion and gathering together of mud and sand; which was the conjecture of Herodotus long before Strabo. In like manner that the Iland Pharos, which in Plinies time was joyned to Alexandria by a bridge, as himselfe testifieth, lib. 5, cap. 31 (and therefore for this reason may seeme to have been called a Peninsula by Strabo), was anciently distant from Egypt a whole day and nights saile, is reported both by Pliny and Solinus out of Homer. And this is the reason, as Strabo conjectures, that Homer (whereas he makes often mention of Thebes in Egypt) yet speakes not one word of Memphis; and that either because at that time it was a very small place, or else perhaps it was not as yet in being, the land being in Homers time covered all over with water where Memphis was afterwards built. And this seemes also to be confirmed by the great depression and lownesse of the intermediate shore betwixt the two Seas, which is so great that when Sesostris first had an intent of cutting a channell betwixt the two Seas, as was afterwards intended also by Darius, and lastly by Ptolomy, they were all forced by this reason to desist from their enterprise. And, indeed, Strabo reports that himselfe saw the Egyptian shore in his time all overflowed beyond the Mountaine Cassius. Besides, the great retiring of the waters at an ebbe, as well in the Arabian gulfe as in the Persian, seemes somewhat to confirme this conjecture of Eratosthenes. For the tides withdraw themselves so farre back in the Arabian gulf that Julius Scaliger makes mention of some cavillers that, for this very reason, went about to derogate from the miraculous passage of the Children of Israel for the space of above 600 miles through the Red Sea, as if they had watched their time when the tide gave way, and that when it returned againe the Egyptians were overtaken therewith and all drowned.[1]

And it is reported by Pliny that Numenius, generall to Antiochus, fighting against the Persians near the mouth of

[1] This sea, says Pontanus, is always rendered Erythræum in the Septuagint, and Rubrum by St. Jerome.

the Persian Gulfe, not far from the promontory called Macavum, got the victory twice in one day, first by a sea combat and afterward (the water having left the place dry) on horsbacke, as is related by him in his 6 booke, 28 cap.

And thus much concerning Eratosthenes his conjecture. Let us now returne to the bounds of Africa, which is divided (as we have already said) on the East from Asia by a Meridian drawne through the Arabian gulfe to the Mediterranean Sea. On all the other sides it is encompassed about with the Sea; as on the West with the Atlanticke; on the South with the Æthiopian Ocean; and on the North by the Mediterranean, which is also the Southerne bound of Europe.

Now as concerning Ptolomies ignorance of the Southerne parts of Africa, making it a continent and contiguous to Asia by a certaine unknowne land, which he would have to encompasse about the South side of the Indian Sea and the Æthiopian gulfe; if it be not sufficiently evinced out of the relations of the Ancients, as namely of Herodotus, who reporteth that certaine men were sent forth by Darius by Sea, who sailed all about this tract; nor yet of Heraclides Ponticus, who relates a story of a certaine Magician who said that he had compassed about all these coasts, because Posidonius accounteth not these relations of credit enough to conclude anything against Polybius; neither doth he approve of that story of one Eudoxus Cyzicenus, reported by Strabo, Pliny, and Mela, out of Cornelius Nepos, an Author of very good esteeme (and that because Strabo thought this relation to deserve no more credit then those fabulous relations of Pytheas, Evemerus, and Antiphanes), nor lastly those traditions of King Juba concerning the same matter related by Solinus. Howsoever, I say that these traditions of the Ancients doe not convince Ptolomy of ignorance; yet certainely the late navigation of the Portugals most evidently demonstrate the same, who, touching upon the most outward point of all Africa, which they now call the Cape of Good

Hope, passe on as farre as the East Indies. I shall not in the meane time neede to speake at all of that other story which Pliny hath, that at what time C. Cæsar, sonne to Augustus, was proconsul in Arabia, there were certaine Ensignes found in the Arabian gulfe which were knowne to be some of those that were cast away in a shipwracke of the Spanish Navy; and that Carthage being at that time in her height of power, Hanno, a Carthaginian, sailed about from Gades as farre as Arabia, who also afterward himselfe wrote the story of that navigation.

Asia lyeth Eastward both from Europe and Africa, and is divided from them by these bounds and limits which we have already set downe. In all other parts it is kept in by the Ocean. On the North by the Hyperborean or Frozen Sea; on the East by the Tartarian and Easterne Ocean; on the South by the Indian and Red Sea. But Ptolomy would have the Northerne parts of Asia, as also of Europe, to be encompassed not with any Sea, but with a certaine unknowne land; which is still the opinion of some of our later writers, who think that country which we call Greenland to be a part of the Indian Continent. But we have very good reason to suspect the truth of this their opinion; since that so many Sea-voyages of our own country-men, who have gone farre within the Arcticke circle, beyond the utmost parts of Norway, and into that cold frozen Channell that divides Nova Zembla from Russia, doe sufficiently testifie that all those parts are encompassed by the Sea. Not to speake anything of that which Mela alleadgeth out of Cornelius Nepos, how that when Q. Metellus Celer was Proconsul in Gallia there were presented him by the King of Suevia certaine Indians, who having beene severed by force of tempests from the Indian shore, had been brought about, by the violence of windes, as farre as Germany. Neither will I here mention that other relation of Patrocles in Strabo, who affirmed that it was possible to saile to India all along the

Sea shore a great deale more Northward than the Bactrians, Hircania, and the Caspian Sea. Now Patrocles was made governour of these places. Nor lastly that which Pliny himselfe reporteth, how that all this Eastern coast, from India as farre as to the Caspian Sea, was sailed through by the Macedonian Armies in the raigne of Seleucus and Antiochus.

Concerning the quantity of the Earth which was inhabited, there was great diversity of opinion among the Ancients. Ptolomy defined the longitude of it to be, from West to East, beginning at the Meridian which passeth through the Fortunate Islands, and ending at that which is drawne through the Metropolis of the Sinæ or Chineans countrey. So that it should containe halfe the Æquator, which is 180 degrees and 12 Æquinoctiall houres, or 90,000 furlongs measured by the Æquator. And he determined the bounds of the latitude to be, toward the South, that Parallel which lyeth 16 gr. 25 m. Southward of the Æquator; and the Northerne limits to be made that Parallel which passeth through Thule or Iseland, being distant from the Æquinoctial 63 degrees. So that the whole latitude of it contained in all 79 gr. 25 m., or 80 whole degrees, which is neare upon 40,000 furlongs. The extent of it, therefore, from East to West, is longer then it is from North to South, under the Æquinoctiall something more then by halfe as much, and under the most Northerne Parallel almost by a fiftieth part. Good reason, therefore, had the Ancient Geographers, as Ptolomy in his lib. 1, Cap. 6, Geograph., to call the extent of it from West to East the Longitude of it, and from North to South the Latitude. Strabo also acknowledgeth the Latitude with Ptolomy to be 180 degrees in the Æquator, as likewise Hipparchus doth also; notwithstanding there is some difference betwixt them in the number of the furlongs. For these last have set downe the Longitude to be 126,000 furlongs under the Æquator: herein following Eratosthenes, who reckoneth 700 furlongs to a degree. But Strabo maketh the latitude a great deal lesse, that is, something lesse then 30,000 furlongs; and hee boundeth it on the South with the Parallel drawne through Cinnamomifera, which is distant Northward from the Æquator 8800 furlongs, and on the North with that Parallel which passeth through those parts, which are 4000 furlongs or thereabouts more Northward then Britaine. And this Parallel that passeth through the Region called Cinnamomifera, Strabo makes to be more Southward then Taprobane, or at least to pass through the most Southerne parts of the same. But herein he betrayeth his owne notable ignorance, for as much as the most Southerne part of this Iland is extended farre beyond the Æquator; as both Ptolomy affirmeth in his Geography, lib. 7, Cap. 4, and is further confirmed by the late Navigations of the Portugals. But Dionysius Afer is much farther out of the way than so, for he placeth Taprobane under the tropicke of Cancer.

And these were the bounds wherewith the Ancient Geographers terminated the then inhabited parts of the World. But in these riper times of ours, by the industry at Sea both of the Spaniards, English, and others, the Maritime coasts of Africa have beene more thoroughly discovered, to above 35 gr. of Southerne Latitude; and the Northerne limits of Europe have now been searched into as farre as the 73 degree of Northerne Latitude, farre within the Articke circle; besides all that which hath at length beene discovered in the New World, beyond the hope or opinion of any of the Ancients, the name of it being not so much as knowne to them.

America, which for its spaciousnesse may well be called the other World, extending itselfe beyond 52 gr. of Southerne Latitude, is there bounded with the Straits of Magellane; and toward the North it runneth farre within the Arcticke circle; on which side also that it is bounded by the Sea, the many Navigations of our Countrey-men into those parts doe give strong arguments of hope. I shal not here speak of those Sea coasts which are beyond that Sea that encompasseth about the most Northerne parts of Europe and Asia, as having beene but only seene afarre off as yet, and not throughly discovered. Nor yet those other which are more Southerne then the Indian and Red Seas; which as yet we have not any experience to the contrary, but that wee may beleeve to bee one continent with those other Southerne Lands that lye beyond the Straits of Magellane.

Europe (whether so called from Europa Tyria, daughter to Agenor, as some thinke; or Phœnix, as Herodotus will have it; or else from Europa, a Sea Nymph, according to the opinion of Hippias in Eustathius; or else from Europus, as Nicias in the same Eustathius would have it to be) containeth in it these principal regions, to wit, Spaine, France, Italy, Germany, Bohemia, Prussia, Rhœtia, Livonia, Sclavonia, Greece, Hungary, Polonia, Moscovia or Russia, Norway, Sweden, and Denmarke. To these wee may add the principall Islands, as namely those of Great Britaine, the chief of which is England and Scotland, ennobled chiefly by being united to the English Crowne; as also Ireland, which is in like manner subject to the same. Besides the Azores and many other Islands scattered up and downe in the Mediterranean Sea, as Sicily, Sardinia, Crete, etc.

Africa (whether it be so called from Apher, one of Hercules his companions in his expedition against Gerion, according to Eustathius; or else from one Iphricus, a certaine king of the Arabians, whence also it is called in Arabique Iphricia, as Johannes Leo testifieth; or lastly from its scorching heat, as if it should be called *αφρικη*, *quasi sine frigore*, as some are pleased to derive it) hath in it these principall regions. First of all, next to the Straits of Gibraltar (anciently called Fretum Gaditanum) there lyeth Barbary, heretofore called Mauritania, which containeth in it the kingdomes of Morocco, Fez, Tunis, and Algier. Next to

Barbary lyeth Egypt, which also bordereth upon the Mediterranean Sea. Now within Barbary toward the continent there lyeth Biledulgerid, known to the Ancients by the name of Numidia. The 3d is that part which is called by the Greekes and Latines Libya; but the Arabians name it Sarra. After this followes the countrey of the Negroes, so called because they border upon the river Niger, or else from their colour. This countrey is now called Senega, and it hath in it many petty kingdomes, as, namely, Gualata, Guinea, Melli, Tombutum, Gagos, Guberis, Agodes, Canos, Casena, Zegzega, Zanfaran, Burnum, Gaoga, Nubia, etc. Next to these is the spacious territory of the King of the Æthiopians (who is also called Pretegiani, and corruptly Prester John), which kingdome is famous for the long continuance of the Christian Religion in it, which hath been kept amongst them in a continuall succession ever since the Apostles time. These Christians are called Abyssines, but more rightly Habassines, as Arius Montanus observeth in the itinerary of Benjamin Tudelensis. Their dominion was anciently extended very farre through Asia also. These have bordering on the West some few obscure kingdomes, as Manicongo and D'Angola; and toward the East and South, Melinde, Quiloa, Mozambique, Benamatapa. The chiefe Islands that are situate neare it are Madagascur, the Canary Islands, the Isles of Cape Verd, and St. Thomas Island, lying direct under the Æquator.

Asia (so called from Asia, the mother of Prometheus, as the common received opinion is; or else from a certaine Hero of that name, as Hippias in Eustathius wil have it), at this day wholly in subjection to the Great Turke and the Persians as farre as to the East Indies, the greatest part whereof is under the kings of China and Pegu. But the more Northerne parts of Asia are possessed by the Muscovites, Tartarians, and those that inhabit the regions of Cathaia. The principall Islands appertaining unto it are Cyprus and

Rhodes in the Mediterranean; and on the South side Sumatra, Zeilam, Java Major and Minor, the Moluccan and Philippine Islands, besides Borneo, and almost an infinite company of others. And on the East of it there lye the Japonian Islands.

America (so called from Americus Vespuccius, who first discovering it, gave it both name and bounds) is terminated on the East side (on which it lookes toward Europe and Africa) by the Atlanticke Ocean; on the West with the Sea which they call del Zur, or the South Sea; on the South it is bounded with the Straits of Magellane. But as for the Northerne parts of it, they are not yet thoroughly discovered, or the limits thereof knowne, notwithstanding many adventures by Sea of our Countrymen, Mr. Martin Frobisher and Mr. John Davis, have given strong arguments of hope that it is on that side bounded by the frozen Sea. It containeth in it these principall regions. First on the North, that country which the Spaniards call Tierra de Labrador. After which followeth that which they call Baccalearum Regio, then Nova Francia, after this Virginia, then Florida. Next to this Nova Hispania, famous especially for the City of Mexico; and last of all the kingdomes of Brazilia and Peru, which are the most Southerne parts of all. There are also many adiacent Islands, most of which lye in the Bay of Mexico, eastward from America; the most notable of which are Cuba and Hispaniola, besides many others of lesse note.

Terra incognita.

There are also many other parts of the world not yet thoroughly knowne or discovered, as, namely, those Southerne coasts wherein stands Nova Guinea, lying beyond the Indian Sea, which, whether it be an Island or part of the Maine Continent, is not yet discovered; and likewise that other tract of the Southerne unknowne Continent which is called Magellanica; as also those Northerne parts of Europe, Asia, and America which have beene but lately detected by many of our English Navigators, but not as yet fully searched into.

CHAPTER II.

Of the Circumference of the Earth, or of a Greater Circle; and of the Measure of a Degree.

It remaineth now that we speake somewhat of the circumference of the Earth, or of the greatest Circle in it, the knowledge whereof is very necessary, both for the study of Geography as also for the easier attaining to the Art of Navigation. And therefore I hope I shall not seeme impertinent, if I insist something the longer on this argument, especially seeing that there is great diversity of opinion among the most learned Authors that are extant, concerning this matter; insomuch that it is not yet determined which of them we are to follow.

De ambitu terra.

Arist

Aristotle, in the end of his 2d booke, de Cœlos, affirmes (and that according to the doctrine of the Mathematicians, as himselfe saith) that the circumference of the Earth is 400,000 furlongs. Cleomedes, lib. 1, reckons it to be 300,000, for he saith that the Vertical Points of Lysimachia and Syene were observed by Sciotericall Instruments to be distant from each other the 15th part of the same Meridian. Now the distance between these two places hee sets downe to be 20,000 furlongs. So that if 20,000 be multiplied by 15, the whole will arise to 300,000. Eratosthenes (if we may beleeve Strabo, Vitruvius, Pliny, and Censorinus) would have the whole compasse of the Earth to containe 252,000 furlongs. To which number Hipparchus, as Pliny testifieth, added very near 25,000 more. Yet Strabo, as well in the end of his 2d booke of his Geography, as elsewhere, affirmeth that he used the same measure that Eratosthenes did; where he saith that, according to the opinion of Hipparchus, the whole quantity of the Earth containeth 252,000 furlongs; which was the measure delivered also by Eratosthenes.

Cleom.

Strabo passim. Vitr., lib. 1, c. 6. Plin.. lib. 2, c. 108. Censor., c. 13.

Which opinion of Eratosthenes is seconded also by that fabulous relation of Dionysiodorus, recorded by Pliny, lib. 2, Cap. ult., where he saith that there was found in the sepulchre of Dionysiodorus an epistle written to the Gods; wherein was testified that the semidiameter of the Earth contains 4200 furlongs, which number being multiplied by 6 the product will bee 252,000.

Cleomedes, relating the observations of Eratosthenes, and Cleom.. l. 2.
Posidonius maketh it to be somewhat lesse, and that according to the doctrine of Eratosthenes, to wit, 250,000 furlongs. For he placeth Alexandria and Syene under the same Meridian. Now Syene being situate direct under the Tropicke, the Sunne being then in the Summer Solstice, the gnomons cast no shadow at all. For confirmation of which, the experiment was made by digging a deepe well, which at that time of the yeare was wholly enlightened on every part, as it is reported both by Pliny, and also by Strabo before him. But at Alexandria, when the Sunne is in the Summer Tropicke, the gnomon is observed to cast a shadow to the fiftieth part of the circumference, on which it is erected to right angles, so that the top of the same is the center of the circumference. Now the distance between Syene and Alexandria, is commonly set downe by Eratosthenes, Pliny, and Strabo to be Lib. 2, c. 73
5000 furlongs. If, therefore, 5000 be multiplied by 50, the whole will arise to 250,000, which is the number of furlongs assigned to the circumference of the whole earth by Eratosthenes. Posidonius, proceeding by another method, though not unlike this, labours to prove the whole circuit of the Earth to containe 240,000 furlongs. And first hee taketh for granted (which is also acknowledged by Ptolomy, lib. 5, cap. 3. Almagest) that Rhode and Alexandria are situate under the same Meridian. Now that bright Star in the sterne of Argo (which they call Canobus, and which never appeareth in Greece, which seemes to be the reason why Aratus maketh no mention of it), first beginneth to appeare

above the Horizon at Rhodes; but it doth but *stringere Horizontem,* just touch the Horizon, and so upon the least circumvolution of the Heavens setteth againe, or else, as L. de Sphæ. Proclus saith, is very hardly seene unlesse it be from some eminent place. But when you are at Alexandria you may see it very cleare above the Horizon. For when it is in the Meridian, that is at the highest elevation above the Horizon, it is elevated above the Horizon about the fourth part of a Signe; that is to say the forty eighth part of the Meridian that passeth through Rhodes and Alexandria. The same is affirmed also by Proclus, if you read him thus: "Canobum in Alexandria conspicue cerni quarta circiter Signi portione supra Horizontem extante", as it ought to be, and not as it is corruptly read in Alexandria, "*prorsus non cerni.*" "It is not seene at all", instead of: "It is seene very plainely", *αφανης* being crept into the text perhaps instead of *ευφανης*. Now the distance betwixt Rhodes and Alexandria is set L. 2, c. 70. downe both by him and Pliny to be 5,000 furlongs, which being multiplied by forty-eight, the product will be 240,000, the number of furlongs agreeing to the measure of the Earths circumference, according to the opinion of Posidonius.

Ptolomy everywhere in his Geography, as also Marinus Tyrius before him, have allowed but 500 furlongs to a degree in the greatest circle on the earth, of which the whole circumference containeth 360, so that the whole compasse of the Earth, after this account, containeth but 180,000 furlongs. And yet Strabo affirmeth in his lib. 2, Geograph., that this measure of the Earths circumference set downe by Ptolomy was both received by the Ancients, and also approved by Posidonius himselfe.

Strabo, pa. 65. So great is the difference of opinions concerning the compasse of the earth; and yet is every one of these opinions grounded on the authority of great men. In this so great diversity therefore it is doubtfull whom we should follow.

And if you should desire to know the cause of all these dissensions, even that also is altogether as uncertaine. Nonius and Pucerus would perswade us that certainely the furlongs they used were not of the same quantity. Maurolycus and Philander conceive the difference of furlongs to rise out of the diverse measure of Pases. And therefore Maurolycus takes great paines to reconcile them; but in vaine, for they seeme not capable of any reconcilement. They tell us of diverse kinds of Pases among the Ancients. It is true; wee assent to them herein; but withal desire to hear of some diversitie of furlongs also, or at least of feet. The Greekes (as I conceive) measured not their furlongs by Pases, but by feet, or rather *ταις οργμαις*. Now *οργμα* is the measure of the extension of both hands, together with the breast betwixt, containing six feet, which we commonly call a fadome, and is a measure in continual use with our Mariners in sounding the depth of the sea or other waters. The word, notwithstanding, is translated by many a Pase, but how rightly I leave it to learned men to judge. Xylander, in his translation of Strabo, alwayes rendereth it an Ell. In like manner a furlong is defined by Herodotus, a very Ancient Greeke Author, to consist of 600 feet; the same also is affirmed by Suidas, by much later than hee. Yet Hero Mechanicus (or at the least his Scholiast, one as I conceive of the lowest ranke of Ancient Writers), will have a furlong to containe 100 fadomes; a fadome foure cubits; a cubit a foote and a halfe, or twenty foure digits. But you will say, perhaps, that Censorinus proposeth three severall kindes of furlongs; the first of which is the Italian, consisting of 625 feet, which he would have us to understand to be that which is commonly used in measuring the Earth. The second is the Olympian, containing 600 feet; and the third and last is the Pythean, consisting of 1,000 feet. But to let passe this later, if wee doe but looke more nearly into the matter we shall find the Julien and Olympian furlongs, however they differ in names,

Nonius de Crepuscul.
Puce. de dim. terræ.
Maur. dialog. 3.
Cosmogr. Phil. in Vitr.

yet to be no other but the selfe same thing. For the Italian furlong, which containeth 625 Romane feete (according as Pliny testifieth in his second booke and twentie third Chapter), will be found to be equall to the Olympian, consisting of 600 Grecian feete. For 600 Grecian feet are equall to 625 Romane; for as such as the Grecian foote exceeds the Roman by a twenty-fourth part, as much is the difference betwixt 600 and 625.

Amongst these so great diversities of opinions, let us give our conjecture also, both what may be the cause of so great disagreement, and also which of them we may most safely follow. We will therefore pass by Aristotle, whose assertion is only defended by a great name. And for Cleomedes his opinion of the earths being in compasse 200,000 furlongs, we should scarce vouchsafe to mention it, but that Archimedes also had taken notice of the same, as of a position not altogether disallowed in his time. Let us therefore examine Eratosthenes and Posidonius, whose opinions seeme to be grounded on more certaine foundations. The cause therefore of their disagreement I conceive to bee in that neither of them had measured exactly the distances of those places which they layd downe to work on, but tooke them on trust from the common received report of Travailers; save only that of the two, Posidonius is the more extravagant. Whereas on the contrary Ptolomy grounded his opinion on the distances of places exactly measured, as himselfe affirmeth, when he saith that the latitude of the knowne parts of the world is 79 degrees, 45 minutes. Or supposing it to be full 80 degrees, it will then containe 40,000 furlongs, allowing for every degree five hundred furlongs; as by measuring the distances of places exactly wee have found it to be.

Eratost. et Posid. executiuntur.

But Eratosthenes is much taxed by Hipparchus for his strange mistakes and grosse ignorance in setting downe the distances of places, as Strabo testifieth in his first booke. For hee reckons betwixt Alexandria and Carthage above 13,000 furlongs, whereas, saith Strabo it is not above 9000. So likewise Posidonius is to bee blamed for setting downe the distance betwixt Rhodes and Alexandria to bee 5000 furlongs, and that from the relations of Mariners, whereas some of them would have it to bee but 4000 and others 5000, as Eratosthenes confesseth in Strabo ; but addeth moreover that he himselfe had found by sciotericall instruments, that it was but 3,750. And Strabo would have it to bee something lesse than that, namely, 3,640 furlongs. So that hence wee may safely conclude that Ptolomies opinion being grounded upon the more exact and accurate dimensions of distances (as himselfe professeth), must necessarily come nearer the truth then the rest.

But Franciscus Maurolycus, Abbot of Messava, while he goes about to defend Posidonius against Ptolomy, is overtaken himselfe in an errour, before hee is aware. For he suspecteth the truth of Ptolomies assignement of the latitude of Rhodes, which he sets downe to be thirty-sixe degrees, and hee advertiseth us, that certainely the numbers in his geographicall tables are corrupted, which we confesse is most certaine. But in the meane time let us see how he proves them to be so in this latitude of Rhodes. Posidonius (saith he) out of his owne observations, setteth downe the latitude of it to be thirtie-eight degrees and an halfe ; unlesse that Ptolomy bee out also in designing the latitude of Alexandria, which Maurolycus thinks cannot possibly be. But we affirme on the contrary side that Ptolomy himselfe is against the latitude, not only in his Geographicall bookes, but also in diverse places throughout the Almagest also, and especially in the lib. 2, cap. 6, where he sets downe the same latitude for Rhodes that he hath in his Geography ; adding moreover the quantity of the longest day, and also what manner of shadowes the gnomons cast, both when the Sun is in the Æquinoctiale, as also in the Tropicke, all which doe plainly prove the same. He also very often hath

Ptol. contra Posidonium defensus. Maurolycus taratur.

the same latitude of it in his Planisphære; unlesse you will say that either Masses the Arabian, in translating it into Arabique, or else Rudolphus Brugensis, who translated the same againe out of Arabicke into Latine, have deceived us. Hitherto therefore wee stand on equall tearmes. But he proceeds and saith that this opinion of Posidonius is favoured also by Proclus, and the observations of Eudoxus Cnidus delivered by Strabo. Let us therefore see what all this is. Posidonius (saith Strabo) reports that himselfe being sometime in a city distant from the Gaditane Straits 400 furlongs, saw from the top of an high house a certaine Starre, which hee tooke to bee Canobus, and those that went thence more southward from Spaine confesse that they saw it also plainely. Now the Tower Cnidus, out of which Eudoxus is said to have seene Canobus, is not much higher than the other buildings. But Cnidus is on the same Climate with Rhodes, as is also the Gades, with the sea coasts adjoyning. Thus Strabo.

Strabo, l. 2, p. 2.

But what doth he conclude hence against Ptolomy? That Canobus may be seene in Cnidus? Wee deny it not. Or that Cnidus is in the Rhodian Climate? Ptolomy acknowledgeth as much, for hee makes it to have not above 39 gr. 15 m. of latitude, in the fifth booke of his Geography. But is not Ptolomy out also in assigning the latitude of Cnidus? That the latitude of Rhodes is no greater than Ptolomy hath set it, may be proved even out of Proclus himselfe; for hee makes the longest day at Rhodes to be fourteene houres and an halfe. And Ptolomy will have the same to be equall both at Rhodes and at Cnidus. And to this assenteth Strabo likewise, save onely that in one place he sets it downe to be but fourteen houres bare; so that by this reckoning it should have lesse latitude. Now Proclus his words are these. In the Horizon of Rhodes (saith hee) the Summer Tropicke is divided by the Horizon, in such sort as that if the whole circle bee divided into forty-eight parts,

Proclus Sphæra.

twenty-nine of the same doe appeare above the Horizon and nineteen lye hid under the Earth. Out of which division it followes that the longest day at Rhodes must be fourteen Æquinoctiall hours and an halfe, and the shortest night nine and a half, thus hee saithe. I do not deny, but that Posidonius, his setting downe of the quantity of the portion of the Meridian intercepted betwixt the verticall point of Rhodes and Alexandria, might deceive Pliny, Proclus, and others. Yet Alfraganus draweth his second Climate through Cyprus and Rhodes, and maketh it to have the longest day of fourteen houres and an halfe, and in latitude 36 gr. two-thirds. So that here is very little difference betwixt him and Ptolomy. And even Maurolycus himselfe, when in his Cosmographicall Dialogues he numbereth up the Parallels, maketh that which passeth through Rhodes to have 36 gr. and a twelfth of latitude; herein differing, something with the most, from Posidonius. Eratosthenes his observations also doe very much contradict Posidonius. For Eratosthenes saith that hee found by sciotericall gnomons, that the distance betwixt Rhodes and Alexandria was 3750 furlongs. But let us examine this a little better. The difference of Latitude betwixt these two places he found sciotericallly, after his manner, to be something more than 5 degrees. And to this difference (according to his assumed measure of the compasse of the Earth, wherein he allows 700 furlongs to a degree) he attributes 3650 furlongs. Neither is there any other way of working by sciotericall instruments (that I know) in finding out the distance of furlongs betwixt two places; unlesse we first know the number of furlongs agreeing either to the whole circumference of the Earth, or else to the part of it assigned. Let us now see if we can prove out of the observations of Eratosthenes himselfe, that neither Posidonius, his opinion concerning the measure of the Earths circumference, much lesse Eratosthenes his owne can be defended. And here we shall not examine his observation of the difference of latitude betwixt Alexandria and Syene, that so we might prove out of his own assumption that the whole compasse of the Earth cannot be above 241,610 furlongs, as it is demonstrated by Petrus Nonius, in his lib. 2, cap. 18, *De Navigatione.* Neither doe we enquire, how truly hee hath set downe the distance of the places to be 5000 furlongs; whereas Solinus reckoneth not from the very Ocean to Meroe, above 620 miles, which are but 4960 furlongs. Now Meroe is a great deal farther than Syene. Neither will we question him at all, concerning the small difference that is betwixt him and Pliny, who reckons from the Island Elephantina (which is 3 miles below the last Cataract, and 16 miles above Syene) to Alexandria, but 486 miles; so that by this reckoning betwixt Syene and Alexandria, there will not be above 4560 furlongs. But we will proceed a contrary way to prove our assertion. This one thing, therefore, we require to be granted us; Which is, that looke how great a space the Sunne Diameter taketh up in his Orbe, for the like space on the Terrestriall Globe shall the Gnomons be without any shadow at all, while the Sunne is in their Zenith. Which if it be granted (as it is freely confessed by Posidonius in Cleomedes) we have then gotten the victory.

Now it is affirmed by Eratosthenes that the Sunne being in the beginning of Cancer, and so directly in the verticall point at Syene; both there and for 400 furlongs round about the gnomons cast no shadow at all. Let us now therefore, see how great a part of his orbe the Sunnes diameter doth subtend. For by this meanes if this position of Eratosthenes, which wee have now set downe, bee true; we may easily finde out by it the whole circuit of

Fir. Mater. the Earth. Firmicus Maternus makes the diameter both of the Sun and Moone to be no lesse then a whole degree. But he is too farre from the truth, and assigneth a greater quantity, either than hee ought or wee desire. The Egyptians

found by hydroscopicall instruments that the diameter of the Sunne takes up the seven hundred and fiftieth part of his Orbe. So that if 300 furlongs on Earth, answer to the seven hundred and fiftieth part of the whole circumference of the same, the whole circuit of it then will be but 225,000 furlongs. The fabricke and use of this instrument is set downe by Proclus in his cap. 3, *Designation. Astronomi.* And Theon also speaks much of it in his Commentaries upon the 5 lib., Almagest Ptolom., as also c. 13.
does Maurolycus in his third Dialog. Cosmograph. But these kindes of observation are not approved of by Ptolomy. And Theon also, and Proclus demonstrate them to bee obnoxious to much errour. And therefore we examine the matter yet a little further.

Aristarchus Samius (as he is cited by Archimedes) affirmed that the Sunnes apparent diameter taketh up the seven hundred and fiftieth part of the Zodiaque, that is to say 30 minutes, and is equall to the apparent diameter of the Moone; as he hath it (as I remember) in the 7 and 8 propositions of his booke *De Magnitud. et distant. Solis et Lunæ.* The same was the opinion also of Archimedes himselfe. But in the meane time I cannot free myselfe of a certaine scruple cast in my way by another supposition of the same Aristarchus in the very same booke, where hee would have the diameter of the Moone to bee 2 degrees. Archimedes also, out of his owne observations by dioptricall instruments, hath defined the Suns diameter to bee greater then the 200th part of a right angle, that is to say 27 minutes, yet lesse then the 164th part of a right angle, which is 33 minutes. But he himselfe confesseth that there is no great credit to be given to such like observations as are made by these dioptricall instruments, as by them to bee able exactly to find out the diameter of the Sunne or Moone, seeing that neither the sight nor the hand, nor yet the instruments themselves, by which the observations are to be made, can be every way so exact and sure as not to faile. Ptolomy, by the same dioptricall instruments, as also by the manner of Eclipses, found the diameter of the Sun to containe 30 min. 20 sec., and to be equall to the apparent diameter of the Moone when she is at the greatest distance from the Earth, which is at the full Moone, and in conjunction with the Sunne. Nor whereas he would have this magnitude to bee constantly the same, and invariable: Proclus approves not of him herein, as appeares in the 3 Cap. *Designation. Astronom.*, being hereto induced by the authority of Sosigenes, a Peripatetic, who in these bookes of his which he entituleth, *De revolutionibus,* hath observed in the Eclipses of the Sun there is sometimes a certaine little ring or circle of the Sun to be perceived enlightened, and appearing plainely on all sides round about the body of the Moone. Which if it be true, it is impossible then that the apparent magnitude of the Sunne should be at all times equall to that of the Moone in their conjunctions and oppositions. And this is the cause, perhaps, that those that have come after Ptolomy have endeavoured to examine these things more accurately.

And first of all Albateni found the diameter of the Sunne, when he was in the Apogæum of his Eccentricke, to be 31 min. 20 sec., which is the same with Ptolomies observation; but in the Perigæum to be 33 min. 40 sec. But Copernicus went yet further, and found the diameter of the Sunne, when he was in his greatest distance from the Earth, to be 31 min. 48 sec., and when he is nearest of all, to be 33 min. 54 sec. Now if we worke upon this ground here laid before us, and take the diameter to be 32 min., it will then follow that if 300 furlongs answer to 32 min., the whole circuit of the Earth will bee but 202,500 furlongs: which falls short of that measure which Posidonius hath set downe, but much more of that which Eratosthenes hath delivered. And thus much have we thought good to say (with all due reverence to the judgments of learned Authors) in examination of those things which have been delivered by the Greekes concerning the measure of the Earths circumference.

The way of measuring used here with us is by Miles and Leagues; of the former whereof 60, and of the latter 20 answering to a degree. So that the circumference of the Earth containeth 21,600 English Miles, which also agrees exactly with that of Ptolomy. For we find our English foot to be just equall with the Grecian, by comparing it with the Grecian foot, which Agricola and others have delivered unto us out of their monuments of antiquity. Now one of our Miles containeth 5000 feet of our English measure, and a furlong 600 Grecian feet. Now if you multiply the measure of a furlong by 500 (for so many furlongs doth Ptolomy allot to a degree), and so likewise the measure of a Mile, which is 5000 feet by 60, which is also the number of miles that we reckon to a degree, they will both produce the same number of feet, viz., 300,000. So that from these grounds we may safely conclude that the common computation received among our Mariners doth agree most exactly with that of Ptolomy.

The Italians also make 60 miles to be the measure of a degree; but their measure is something less than Ptolomies. The Germans reckon 15 miles to a degree; one of their Miles containing 4 Italian, so that this reckoning of theirs falls just as much short of Ptolomies as the Italian doth; for according to their computation, a degree containeth not above 480 furlongs, every Italian Mile consisting but of 8 furlongs (unlesse perhaps you rather approve of Polybius his opinion, who (as he is cited by Strabo) over and above 8 furlongs will have 2 Plethra, which is the third part of a furlong, to be added to every mile, which is the just measure of our English Mile). Yet Appian saith that 15 Germane Miles are as much as 60 Italian; and 60 Italian Miles containe 480 furlongs, which is less than Ptolomies measure by 20 furlongs, which make up two Italian miles and an halfe.

Appian in Cosmog.

The Spaniards reckon to a degree, some of them 16 leagues and two third parts, and some seventeene and a halfe. But how their measure stands, compared with the Grecian furlongs, or with the English, Italian, or Germane miles, I have not yet certainely learned. Yet Nonius seemeth to equall the Spanish league with the Schœnus or Parasanga, which if it be so, then those that allow 16 leagues and 2 thirds to a degree have the same measure that Ptolomy hath delivered; but those that allowe 17 and an halfe make it somewhat too large.

Cap. 2, lib. 1, De Navigatione.

It only now remaineth to see what is the doctrine of the Arabians concerning this matter. Of which the most ancient have assigned to the whole circumference of the Earth 24,000 miles or 8000 Parasangæ, so that after this computation a degree must containe 66 miles with two third Parts. And this measure is used by Alhazenus in the end of his booke *de Crepusculis*. Alfraganus, and some of the later Arabicke writers since Almamons time, do generally account 20,400 miles to be the just measure of the Terrestriall Globe. So that one degree containeth by this reckoning 56 miles and a third part. And it is reported by Abulfeda, in the beginning of his Geography, how that by the command of Almamon, King of the Arabians, or Caliph of Babylon, there were certaine men employed who should observe in the plaine field of Singar and the adjoyning sea coasts (meaning the places in a direct line toward the Pole) how many miles answered to a degree; and that they found by a just computation, that in going the space of one degree there were spent full 56 miles without any fractions, and sometime 56 miles and a third part, which make up 1333 cubits with two thirds. But now what proportion the Arabian mile beareth to ours, or the Italian or Germane mile, is not so easie to determine. Yet I conjecture it cannot be lesse than tenne furlongs. The Parasangæ, as Jacob Christmannus tells us out of Abulfeda, that great Arabian Geographer, containeth three Arabian

Christ. in Alfrag.

miles, according to the doctrine of the ancient and moderne writers among them. Now a Parasanga (as it appeareth plainely out of Herodotus, Xenophon, and others) containeth thirtie furlongs; so that by this account every mile must comprehend tenne furlongs. And for confirmation of this we may observe that among the Greekes there were two kindes of cubits in use; the one, the common or ordinary cubit, which contained two foot and an halfe of Grecian measure, or twenty-foure digits, of which sixteene went to a foot. The other was the Kings Cubit, in use among the Persians; which was greater than the common Cubit by three fingers breadth. Now Alfraganus affirmeth that the Arabian mile contained 4000 Cubits according to the ordinary measure. So that if this Cubitt be equall to the Grecian Cubit one of their miles will then containe 6000 Grecian feet, which make up tenne furlongs. Now whereas the Parasanga is reckoned by some to containe 40 furlongs, and by others 60, yet no body alloteth to it lesse then 30. With which later account, if we should with Herodotus, Xenophon, and others, rest ourselves contented, **Agricola.** neither indeed is it our intention to stand long in disputing whether or no in diverse places the measure of the Parasanga was also different, as Strabo seemes to thinke, who observed the very same difference in the Ægyptian Schœnus, when as being conveighed on the River Nilus, from one City to another, he observed that the Egyptians in diverse places used diverse measures of their Schœnus: I say if we should rest upon their determination, who assign but 30 furlongs to a Parasanga, then one of the Arabian miles will containe tenne furlongs at the least. Which conjectures, if they be true, we cannot then assent to those learned men, P. Non- **Nonnius de Crep. 19.** nius and Jacobus Christmannus, who will have the Arabian **Chr. ad 10, 6 Alfrag.** mile to be all one with the Italian.

In this so great diversity of opinions concerning the true measure of the earths circumference, let it be free for every man to follow whomsoever he please. Yet were it not that the later Arabians doe countermand us, by proposing to us their Positions, which they averre to have beene grounded upon most certaine and exact mensurations of the distances of places, we should not doubt to prefer Ptolomies opinion before the rest. And for your better satisfaction I will here propose unto your view a list of all those opinions which carry in them any shew of probability.

	Authors.	*Furlongs.*
The circuit of the whole earth containeth, according to—	Strabo and Hipparchus	252,000
	Eratosthenes	250,000
	Posidonius and the Ancient Arabians	240,000
	Ptolomy and Our Englishmen	180,000
	The Moderne Arabians	204,000
	The Italians and Germans	172,800

	Authors.	*Furlongs.*
The Measure of a degree, according to—	Strabo and Hipparchus	700
	Eratosthenes	$694\frac{4}{9}$
	Posidonius and the Ancient Arabians	$666\frac{2}{3}$
	Ptolomy and our Englishmen	500
	The later Arabian.	$566\frac{2}{3}$
	Italians and Germanes	480

	Miles.		*Furlongs.*
The	Italian	containeth	8
	English	,,	$8\frac{1}{3}$
	Arabian	,,	10
	German	,,	32

THE FOURTH PART.

Of the Use of Globes.

HITHERTO wee have spoken of the Globe itselfe, together with its dimensions, circles, and other instruments necessarily belonging thereto. It remaineth now that we come to the practise of it, and declare its severall uses. And first of all it is very necessary for the practise, both of Astronomy, Geography, and also the Art of Navigation. For by it there is an easie and ready way laid downe, for the finding out both of the place of the Sun, the Longitudes, Latitudes, and Positions of places, the length of dayes and houres; as also for the finding of the Longitude, Latitude, Declination, Ascension both Right and Oblique, the Amplitude of the rising and setting of the Sunne and Starres, together with almost an infinite number of other like things. Of the Chiefe of all which wee intend here briefely to discourse, omitting the enumeration of them all, as being tedious and not suitable to the brevity we intend. Now that all these things may be performed farre more accurately by the helpe of numbers, and the doctrine of Triangles, Plaines, and Sphæricall bodies, is a thing very well knowne to those that are acquainted with the Mathematickes. But this way of proceeding, besides that it is very tedious and prolixe, so likewise doth it require great practise in the Mathematickes.

But the same things may be found out readily and easily by the helpe of the Globe with little or no knowledge of the Mathematickes at all.

CHAPTER I.

How to finde the Longitude, Latitude, Distance, and Angle of Position, or situation of any place expressed in the Terrestriall Globe.

The Ancient Geographers, from Ptolomies time downeward, reckon the longitude of places from the Meridian which passes through the Fortunate Islands; which are the same that are now called the Canary Islands, as the most men doe generally beleeve; but how rightly, I will not stand here to examine. I shall only here advertise the reader by the way that the latitude assigned by Ptolomy to the Fortunate Islands falleth something of the widest of the Canary Islands, and agreeth a great deale nearer with the latitude of those Islands which

Insula de Capo Verde. are knowne by the name of Cabo Verde. For Ptolomy placed all the Fortunate Islands within the 10 gr. 30 m., and the 16 gr. of Northerne latitude. But the Canary Islands

Ferro (W. Pt.), 27, 44. are found to be distant from the Equator at least 27 degrees. The Arabians began to reckon their longitude at that place where the Atlanticke Ocean driveth farthest into the maine land, which place is tenne degrees distant eastward from the Fortunate Islands, as Jacobus Christmannus hath observed out of Abulfeda. Our Moderne Geographers for the most part beginne to reckon the longitude of places from these Canary Islands. Yet some beginne at those Islands which they call Azores; and from these bounds are the longitudes of places to be reckoned in these Globes whereof we speake.

Now the longitude of any place is defined to be an Arch, or portion of the Æquator intercepted betwixt the Meridian of any place assigned and the Meridian that passeth through Saint Michaels Island (which is one of the Azores), or of any other place from whence the longitude of places is wont to be determined.

Now if you desire to know the longitude of any place expressed in the Globe you must apply the same place to the Meridian, and observing at what place the Meridian cutteth the Æquator, reckon the degree of the Æquator from the Meridian of Saint Michael's Island to that place; for so many are the degrees of longitude to the place you looke for.

St. Michael (Delgo de.) 37. 45. 10. N., 25. 41. 80. W.

In the same manner may you measure the difference of longitude betwixt any other two places that are described on the Globe. For the difference of longitude is nothing else but an Arch of the Æquator intercepted betwixt the Meridians of the same Places. Which difference of longitude many have endeavoured to set downe diverse ways how to finde by observation. But the most certaine way of all for this purpose is confessed by all writers to be by Eclipses of the Moone. But now these Eclipses happen but seldome, but are more seldom seene, yet most seldome, and in very few places, observed by the skilfull Artists in this Science. So that there are but few longitudes of places designed out by this meanes.

Orontius Finæus, and Johannes Wernerus before him, conceived that the difference of longitude might be assigned by the known (as they presuppose it) motion of the Moone, and the passing of the same through the Meridian of any place. But this is an uncertaine and ticklish way, and subject to many difficulties. Others have gone other ways to worke; as, namely, by observing the space of the Æquinoctiall houres betwixt the Meridians of two places, which they conceive may be taken by the helpe of sunne dials, or clocks, or houre glasses, either with water or sand, or the like. But all these conceits long since devised, having beene more strictly and accurately examined, have beene disallowed and rejected by all learned men (at least those of riper judgments) as being altogether unable to performe that which is required of them. But yet for all this there are a kind of trifling Impostors that make public sale of these toys or worse, and that with great ostentation and boasting; to the great abuse and expense of some men of good note and quality, who are perhaps better stored with money then either learning and judgment. But I shall not stand here to discover the erroures and uncertaineties of these instruments. Only I admonish these men by the way that they beware of these fellowes, least when their noses are wiped (as we say) of their money, they too late repent them of their ill-bought bargaines. Away with all such trifling, cheating rascals.[1]

CHAPTER II.

How to finde the Latitude of any place.

Latitude quid.

The latitude of a place is the distance of the Zenith, or the verticall point thereof from the Æquator. Now if you desire to finde out the latitude of any place expressed in the Globe, you must apply the same to the Meridian, and reckon the number of degrees that it is distant from the Æquator; for so much is the Latitude of that place. And this also you may observe, that the latitude of every place is alwayes equall to the elevation of the same place. For look how many degrees the verticall point of any place is distant from the Æquator, just so many is the Pole elevated above the Horizon; as you may prove by the Globe if you so order it as that the Zenith of the place be 90 degrees distant every way from the Horizon.[2]

---

[1] Here Pontanus has a note, describing the method of finding the longitude by eclipses of the moon.

[2] Pontanus gives a note here, explaining how to find the latitude by observation of circumpolar stars.

CHAPTER III.

How to find the distance of two places, and angle of position, or situation.

If you set your Globe in such sort as that the Zenith of one of the places be 90 gr. distant every way from the Horizon, and then fasten the quadrant of Altitude to the Verticall point, and so move it up and downe untill it passe through the Vertex of the other place; the number of degrees intercepted in the quadrant betwixt the two places, being resolved into furlongs, miles, or leagues (as you please), will shew the true distance of the places assigned. And the other end of the quadrant that toucheth upon the Horizon will shew on what wind, or quarter of the world, the one place is in respect of the other, or what Angle of Position (as they call it) it hath. For the Angle of Position is that which is comprehended betwixt the Meridian of any place, and a greater circle passing through the Zeniths of any two places assigned; and the quantity of it is to bee numbred in the Horizon. Angulus positionis quid.

As for example, the Longitude of London is twentie sixe degrees, and it hath in Northerne Latitude 51 degrees and a halfe. Now if it be demanded what distance and angle of position it beareth to Saint Michaels Island, which is one of the Azores: we must proceed thus to find it. First, let the North Pole be elevated $51\frac{1}{2}$ degrees, which is the latitude of London. Then, fastning the quadrant of Altitude to the Zenith of it, that is to say, fiftie-one degrees and an halfe Northward from the Æquator, we must turne it about till it passe through Saint Michaels Island, and we shall finde the distance intercepted betwixt these two places to be 11 gr. 40 min., or thereabouts, which is 280 of our leagues. And if we observe in what part of the Horizon the end of the quadrant Exemplum.

resteth, we shall find the Angle of Position to fall neare upon 50 gr. betwixt South west and by west. And this is the situation of this Island in respect of London.

CHAPTER. IV.

To finde the altitude of the Sunne, or other Starre.

Altitude quid.

The Altitude of the Sunne, or other Starre, is the distance of the same, reckoned in a greater Circle, passing the Zenith of any place and the body of the Sunne or Starre. Now that the manner of observing the same is to be performed either by the crosse staffe, quadrant, or other like Instrument, is a thing so well knowne, as that it were vaine to repeat it. Gemma Frisius teacheth a way how to observe the Altitude of the Sunne by a Sphæricall Gnomon. But this way of proceeding is not so well liked, as being subject to many difficulties and errours; as whosoever proveth it shall easily find.

CHAPTER V.

To finde the place and declination of the Sunne for any day given.

Having first learned the day of the moneth, you must looke for the same in the Calendar described on the Horizon of your Globe. Over against which, in the same Horizon, you shall find the Signe of the Zodiaque, and the degree of the same, that the Sunne is in at that time. But if it be leape yeare, then, for the next day after the 28th of February, you must take that degree of the Signe which is ascribed to the day following it. As for example, if you desire to know what degree of the Zodiaque the Sunne is in the 29th of

February, you must take that degree which is assigned for the 1st of March, and for the first of March take the degree of the second, and so forward. Yet I should rather counsell, if the place of the Sunne be accurately to be knowne, that you would have recourse to some Ephemerides where you may have the place of the Sunne exactly calculated for every day in the yeare. Neither indeed can the practise by the Globe in this case bee so accurate as often times it is required to bee.

Now when you have found the place of the Sunne, apply the same to the Meridian, and reckon thereon how many degrees the Sunne is distant from the Æquator, for so many will the degrees be of the Sunne's declination for the day assigned. For the Declination of the Sunne or any other Starre is nothing else but the distance of the same from the Æquator reckoned on the Meridian. But the Sunnes Declination may be much more exactly found out of those tables which Mariners use, in which the Meridian Altitude, or Declination of the Sunne for every day in the yeare, and the quantity of it is expressed. One thing I shall give you notice of by the way, and that is, that you make use of those that are latest made as neare as you can. For all of them, after some certaine space of time, will have their errours. And I give this advertisement the rather for that I have seen some, that having some of these tables that were very ancient, and written out with great care and diligence (which notwithstanding would differ from the later Tables, and indeed from the truth itselfe, oftentimes at least 10 min., and sometimes more), yet would they alwayes use them very constantly, and with a kinde of religion. But these men take a great deale of paines and care to bring upon themselves no small errors.

Quid declinatio.

CHAPTER VI.

How to finde the latitude of any place by observing the Meridian Altitude of the Sunne or other Starre.

Observe the Meridian Altitude of the Sunne with the crosse staffe, quadrant, or other like instrument; and having also found the place of the Sunne in the Eclipticke, apply the same to the Meridian, and so move the Meridian up and downe, through the notches it stands in, untill the place of the Sunne be elevated so many degrees above the Horizon as the Sunnes altitude is. And the Globe standing in this position, the elevation of either of the Poles will show the Latitude of the place wherein you are, an example whereof may bee this.

Exemplum. On the 12th of June, according to the old Julian account, the Sunne is in the first degree of Cancer, and hath his greatest declination 23½ degrees. And on the same day suppose the Meridian Altitude of the Sunne to be 50 degrees, we enquire, therefore, now what is the Latitude of the place where this observation was made? And this wee finde out after this manner. We apply the first degree of the Cancer to the Meridian, which we move up and downe, till the same degree be elevated above the Horizon 50 degrees: which is the Meridian altitude of the Sunne observed. Now in this position of the Globe we find the North Pole to be elevated 63 gr. and an halfe; so that we conclude this to be the latitude of the place where our observation was made.

The like way of proceeding doe Mariners also use for the finding out of the Latitude of places by the Meridian Altitude of the Sunne and their Tables of Declinations. But I shall not here speake any further of this, as well for that the explication thereof doth not so properly concerne our proper intention; as also because it is so well knowne to everybody, as that the handling of it in this place would be needlesse and superfluous.

The like effect may be brought by observing the Meridian Altitude of any other Starre expressed in the Globe. For if you set your Globe, so as that the Starre you meane to observe be so much elevated above the Horizon as the Meridian Altitude of it is observed to be, the elevation of the Pole above the Horizon will shew the Latitude of the place. But here I should advise that the latitude of places bee rather enquired after by the Meridian altitude of the Sunne, then of the fixed Starres ; because the Declinations, as wee have already showed, are very much changed, unlesse they be restored to their proper places by later observations.

Some there are that undertake to performe the same, not only by the Meridian Altitude of the Sunne or Starre, but also by observing it at two severall times, and knowing the space of time or horizontall distance betwixt the two observations. But the practice hereof is prolix and doubtful : besides that, by reason of the multitude of observations that must be made, it is also subject to many errours and difficulties. Notwithstanding, the easiest way of proceeding that I know in this kind is this that followeth.

To finde out the Latitude of any place, by knowing the place of the Sunne or other Starre, and observing the Altitude of it two severall times, with the space of time betwixt the two observations.

First having taken with your Compasses the complement of the Altitude of your first Observation (now the complement of the Altitude is nothing else but the difference of degrees by which the Altitude is found to be lesse then 90 degrees), you must set one of the feet of your Compasses in that degrée of the Ecliptique that the Sunne is in at that time ; and with the other describe a circle upon the super- **Alius modus.**

ficies of the Globe, tending somewhat toward the West, if the observation be taken before noone, but toward the East if it be made in the afternoone. Then having made your second observation, and observed the space of time betwixt it and the former, apply the place of the Sunne to the Meridian, turning the Globe to the East untill that so many degrees of the Æquator have passed by the Meridian, as answer to the space of time that passed betwixt your observations, allowing for every houre fifteene degrees in the Æquator, and marking the place in the Parallel of the Sunnes declination that the Meridian crosseth after this turning about of the Globe. And then setting the foot of your Compasses in this very intersection, describe an Arch of a Circle with the other foot of the Compasse extended to the complement of the second observation, which Arch must cut the former circle. And the common intersection of these two circles will shew the verticall point of the place wherein you are: so that having reckoned the distance of it from the Æquator, you shall presently have the latitude of the same.

The same may be effected, if you take any Starre, and work by it after the same manner; or if you describe two circles mutually crossing each other to the complements of any two Starres.

CHAPTER VII.

How to find the Right and Oblique Ascension of the Sunne and Starres for any Latitude of place and time assigned.

Ascensio et descensio quid.

The Ascension of the Sun or Starres is the degree of the Æquator that riseth with the same above the Horizon. And the Descension of it is the degree of the Æquator that goes under the Horizon with the same. Both these is either Right or Oblique. The Right Ascension or Descension is the degree

Ascensio rectu.

of the Æquator that ascendeth or descendeth with the Sunne or other Starre in a Right Sphære; and the Oblique is the degree that ascendeth or descendeth with the same in an Oblique. The former of these is simple, and of one kind only: because there can be but one position of a Right Sphære. But the later is various and manifold, according to the diverse inclination of the same. Oblique.

Now if you desire to know the Right Ascension and Descension of any Starre for any time and place assigned, apply the same Star to the Meridian of your Globe: and that degree of the Æquator that the Meridian crosseth at the situation of the Globe will shew the Right Ascension and Descension of the same, and also divideth each Hemisphære in the midst at the same time with it.

And if you would know the Oblique Ascension or Descension of any Starre, you must first set the Globe to the latitude of the place, and then place the Starre at the extreme part of the Horizon; and the Horizon will shew in the Æquator the degree Oblique Ascension. And if you turn it about to the West side of the Horizon, the same will also shew in the Æquator the oblique descension of that Starre. In like manner you may find out the Oblique Ascension of the Sunne, or any degree of the Eclipticke, having first found out, in the manner wee have formerly shewed, the place of the Sunne. And hence also may bee found the difference of the Right and Oblique Ascension, whence ariseth the diverse length of dayes.

Exemplum. As for example, the Sunne entreth unto Capricorne on the eleventh day of December, according to the old account. I would now, therefore, know the Right and Oblique Ascension of the degree of the Eclipticke for the latitude of fiftie-two degrees. First, therefore, I apply the first degree of Capricorne to the Meridian, where I find the same to cut the Æquator at 270 gr., which is the degree of the Right Ascension. But if you set the Globe to the latitude of fiftie-two degrees, and apply the same degree of Capricorne to the Horizon, you shall find the 303 gr. 50 min. to rise with the same. So that the difference of the Right Ascension 270 and the Oblique 303 gr. 50 min., will be found to be 33 gr. 50 min.

CHAPTER VIII.

How to finde out the Horizontall difference betwixt the Meridian and the Verticall circle of the Sunne or any other Starre (which they call the Azimuth), for any time or place assigned.

Having first observed the Altitude of the Sunne or Starre that you desire to know, set your Globe to the latitude of the place you are in: which done, turne it about, till the place of the Sunne or Starre, which you have observed, be elevated so much above the Horizon as the Altitude of the same you before observed. Now you shall find that you desire if you take the Quadrant of Altitude, and fasten it to the Verticall point of the place you are in, and so move it together with the place of Sunne or Starre up and downe, untill it fall upon that which you have set downe in your instrument at your observation. Now in this situation of the Quadrant, that end of it that toucheth the Horizon will shew the distance of the Verticall circle in which you have observed the Sunne or Starre to be from the Meridian. As for example.

Exemplum. In the Northerne latitude of 51 gr., on the 11th of March after the old account, at what time the Sunne entreth into Aries, suppose the Altitude of the Sunne before noone to be observed to be thirtie gr. above the Horizon. And it is demanded what is the Azimuth or distance of the Sunne from the Meridian. First, therefore, having set the Globe to the latitude of 51 gr., and fastning the Quadrant of Altitude to the Zenith, I turne the Globe about till I finde the first degree of Aries to be 30 gr above the Horizon. And then the Quadrant of Altitude being also applied to the same degree of Aries, will shew upon the Horizon the Azimuth of the Sunne, or distance of it from the Meridian, to bee about fortie five degrees.

CHAPTER IX.

How to finde the houre of the day, as also the Amplitude, of rising and setting of the Sunne and Starres, for any time or latitude of place.

The Sunne, we see, doth rise and set at severall seasons of the yeare, in diverse parts of the Horizon. But among the rest it hath three more notable places of rising and setting. The first whereof is in the Æquator, and this is called his Æquinoctiall rising and setting. The second is in the Summer Solstice when he is in the Tropique of Cancer, and the third is in the Winter Solstice when hee is in the Tropique of Capricorne. Now the Æquinoctiall rising of the Sun is one and the same in every Climate. For the Æquator alwayes cutteth the Horizon in the same points, which are alwaies just 90 gr. distant on each side from the Meridian. But the rest are variable, and change according to the diverse inclination of the Sphære, and therefore the houres are unequall also.

Now if you desire to know the houre, or distance of time, betwixt the rising and setting of the Sunne when he is in either of the Solstices, or in any other intermediate place, and that for any time or latitude of place, you shall work thus: First set your Globe to the latitude of your place, then having found out the place of the Sunne for the time assigned, place the same to the Meridian, and withall you must set the point of the Houre Index at the figure twelve in the Houre circle. And having thus done, you must turne about the Globe toward the East part, till the place of the Sunne touch the Horizon; which done, you shall have the Amplitude of the Sunnes rising also in the Æquator, which you must reckon, as we have said, from the East point or place of intersection betwixt the Æquator and Horizon. And then if you but turne the Globe about to the West side of the Horizon, you shall in like manner have the houre of the setting and Occidentall Amplitude.

And if at the same time, and for the same latitude of place, you desire to know the houre and Amplitude of rising and setting, or the greatest elevation of any other Starre expressed in the Globe, you must turne about the Globe (the Index remaining still in the same position and situation of the Index as before) till the said Starre come to the Horizon, either to the East or West. And so shall you have plainely the houre and latitude that the Starre riseth and setteth in, in like manner as you had in the Sunne. And then if you apply the same to the Meridian, you shall also have the Meridian Altitude of the same Starre. An example of the Suns rising and setting may be this:

Exemplum. When the Sunne enters into Taurus (which in our time happens about the eleventh of Aprill, according to the Julian account), I desire to know the houre and Amplitude of the Sunnes rising, for the Northerne latitude of fiftie-one degrees. Now to finde out this, I set my Globe so that the North Pole is elevated above the Horizon fiftie-one degrees. Then I apply the first degree of Taurus to the Meridian, and the Houre Index to the twelfth houre in the Houre circle. Which done, I turn about the Globe toward the East till that the first degree of Taurus touch the Horizon, and then I find that this point toucheth the Horizon about the twentie-fifth degree Northward from the East point. Therefore I conclude that to bee the Amplitude of the Sunne for that day. In the meantime the Index strikes upon halfe an houre after foure; which I take to be the time of the Sunnes rising.

CHAPTER X.

Of the threefold rising and setting of Stars.

Besides the ordinary emersion and depression of the Starres in regard of the Horizon, by reason of the circumvolution of the Heavens, there is also observed a threefold rising and setting of the Starres. The first of these is called in Latine, *Ortus Matutinus sive Cosmicus,* the morning or Cosmicall rising; the second, *Vespertinus sive Acronychus,* the Evening or Achronychall; and the last, *Heliachus vel Solaris,* Heliacal or Solar. The Cosmicall or morning rising of a Starre is when as it riseth above the Horizon together with the Sunne. And the Cosmicall, or morning setting of a Starre, is when it setteth at the Opposite part of Heaven when the Sunne riseth. The Acronychall or Evening rising of a Starre is when it riseth on the Opposite part when the Sunne setteth. And the Acronychall setting of a Starre is when it setteth at the same time with the Sun. The Heliacal rising of a Starre (which you may properly call the emersion of it) is when a Starre that was hid before by the Sunne beams beginneth now to have recovered itselfe out of the same and to appeare. And so likewise the setting of such a Starre (which may also fitly be called the occultation of the same) is, when the Starre by his own proper motion overtaketh any Starre, so that by the brightnesse of his beams it can no more be seene.

Now, as touching the last of these kinds, many authors are of opinion that the fixed Stars of the first magnitude do begin to shew themselves after their emersion out of the

Sunne beames, when they are as yet in the upper Hemisphære, and the Sunne is gone downe twelve degrees under the Horizon. But these of the second magnitude require that the Sunne is depressed 13 gr., and those of the third require fourteene, and of the fourth fifteene, of the fifth sixteene, of the sixth seventeen, and the cloudy and obscure Starres require eighteene degrees of the Suns depression. But Ptolomy hath determined nothing at all in this case, and withall very rightly gives this admonishment, lib. 8, cap. alt., Almag., that it is a very hard matter to set downe any determination thereof. For as he there well noteth, by reason of the unequall disposition of the Air, this distance also of the Sunne for the Occultation and Emersion of the Starres must needs be unequall. And one thing more we have to increase our suspition of the incertainty of this received opinion, and that is that Vitellio requires nineteene degrees of the Suns depression under the Horizon before the Evening twilight be ended. Now that the obscure and cloudy Starres should appeare ever before the twilight be downe I shall very hardly be persuaded to beleeve. Notwithstanding however the truth of the matter may be, we will follow the common opinion.

Now, therefore, if you desire to know at what time of the yeare any Starre riseth or setteth in the Morning or the Evening, in any climate whatsoever, you may find it out thus: First set your Globe to the latitude of the place you are in, and then apply the Starre you enquire after to the Easterne part of the Horizon, and you shall have that degree of the Eclipticke with which the said Starre rises Cosmically and setteth Acronychally; and on the opposite side on the West, the Horizon will shew the degree of the Eclipticke with which the said Starre riseth Acronychally and setteth Cosmically. For the Cosmicall rising and Acronychall setting, and so likewise Acronychall rising and Cosmicall setting of a Starre are all one, according to those old verses:

> "Cosmice descendit signum, quod
> Acronyche surgit
> Chronyche descendit signum, quod
> Cosmice surgit."

But these things are to be explained more fully. For a Starre doth not alwayes rise and set with the same degree of the Eclipticke. For the Southerne Starres doe anticipate the degree with which they rise at their setting; but the Northerne Starres come after it: that is, if the elevation be of the Articke Pole. Otherwise it is quite contrary if the South Pole be elevated. Now having found the degree of the Eclipticke with which the Starre you enquire after doth rise and set, if you seeke for the same degree of the signe in the Horizon of your Globe, you shall presently have the moneth and day expressed wherein the Sunne commeth to the same degree and signe.

And as for the Heliacal rising and setting of a Starre, you may find it thus. Having set your Globe to the latitude of your place, you must turne about the Starre proposed to the West side of the Horizon, and withall on the opposite East part, observe what degree of the Eclipticke is elevated above the Horizon 12, 13, 14, or any other number of degrees that the magnitude of your Starre shall require for distance from the Sunne. And when the Sunne shall be in the Opposite degree to this, then that Star will set Heliacaly, that is to say, it will be quite taken out of our sight by the brightnesse of the Sunne beames. Now, if on the other side you apply the same Starre to the East, and find out the Opposite degree in the Eclipticke on the West part, that is, the same number of degrees above the Horizon when the Sunne commeth to this place, the same Starre will rise Heliacaly, or recover itselfe out of the Sunne beames. And so if you but find the same degrees of the Eclipticke among the Signes on the Horizon of your Globe, you have the moneth and the day when the Sunne will be in those degrees. And the same also is the time of the emersion and occultation of the Starre you enquire after. But we will here propose an example of the occultation of some fixed Starre of the first magnitude, which done, the emersion of the same is also found by the contrary way of working.

Exemplum.

And the Starre we propose shall be that bright Starre in the mouth of the Great Dog, which is called Sirius, whose occultation we desire to know for the latitude of 51 gr. Northward. Now this Starre, being of the first magnitude, beginnes to bee hid when as it toucheth the Horizon in the upper Hemisphære and the Sunne is at the same time depressed under the Horizon but 12 degrees. If, therefore, you apply this Starre to the West part of the Horizon (having first set your Globe to the latitude of 51 degrees), and on the Opposite East side observe what degree of Eclipticke is just 12 degrees above the Horizon (now this degree is very neare the 11 gr. of Scorpius), when the Sunne shall come to the Opposite degree in the Eclipticke, which is the 11 of Taurus, that Starre will set Heliacaly, and be hid by the Sunne beames. But the Sun comes to this degree of Taurus about the 22 of Aprill; therefore we conclude that the Dogge Starre sets Heliacaly about that time. And if you worke in the same manner, applying the Starre to the East part of the Horizon, you shall have the time of its Heliacal rising or emersion out of the Suns beames.[1]

Not unlike this is the manner of proceeding also in finding the beginning and ending of the twilights; of which we shall speake in the next chapter.

---

[1] Pontanus here inserts an interesting note on the references to these kinds of rising and setting of stars, in the Georgics of Virgil.

CHAPTER XI.

How to finde the beginning and end of the Twilight for any time, and Latitude of Place.

The Twilight is defined to bee a kind of imperfect light betwixt the day and the Night, both after the setting and before the rising of the Sunne; of which the first is called Evening Twilight and the other the Morning. Now the beginning of the one, and the ending of the other, are perceived at the same equall space of time from the rising and setting of the Sun: notwithstanding, the continuance of each of them is sometime greater and sometime lesse. For in Summer the Twilights are much longer then in the Winter. The measure of them they commonly make to be, when as the Sunne is depressed, 18 degrees under the Horizon. But, as P. Nonius rightly observeth, there cannot be any certaine measure or tearme assigned to them, by reason of the various disposition of the aire, and the elevation of the vapours that are exhaled out of the earth; which the same Author saith he findes to be also diverse, sometimes higher and sometimes lower. Vitellio, and Alhazenus before him, would have it to bee, when the Sun is depressed under the Horizon, nineteen degrees. But however the truth be, we shall follow the common received opinion herein. Now, therefore, if you desire to know upon these grounds here laid downe, at what houre the Twilight begins and endeth at any time or latitude of place, you must doe thus: First set your Globe to the latitude of that place, and apply that degree of the Eclipticke wherein the Sunne is in at that time to the Meridian, and withall direct the point of the Index to twelve in the Houre circle; then making the degree of the Eclipticke, that is directly opposite to the place of the Sunne, turne about your Globe, till such time as the opposite degree of the Sunne be elevated eighteene gr. above the Horizon toward the West part of it; and forthwith the Index will shew in the Houre circle the beginning of the Morning Twilight. And if you turne about your Globe in like manner to the East, you shall also have the Houre when the Evening Twilight endeth.

CHAPTER XII.

How to find the length of the Artificiall Day or Night, or quantity of the Sunne's Parallel that remaines above the Horizon, and that is hid beneath it, for any Latitude of place and time assigned. As also to find the same of any other Starres.

The day we have already showed to be twofold, either naturall or artificiall. The natural day is defined by the whole revolution of the Æquator, with that portion also of the same that answereth to such an Arch of the Eclipticke which the Sunne passeth over in one day. Now the whole revolution of the Æquator (besides that portion which answereth to the Sunne's proper motion) is divided into twentie foure equall parts, which they call equall houres, because they are all of equall length, fifteene degrees of the Æquator rising, and as many setting every houre's space. Now the beginning of this day being diverse, according to the diversity of countries, some beginning at Sunset, as the Athenians and Jewes, some at midnight, Ægyptians and Romanes; others at Sunne rising, as the Chaldeans; or at Noone, as the Umbrians, and commonly our Astronomers doe at this day; this being not a thing suitable to our present purpose, I shall not proceed any further in the explanation of the same.

The artificiall day is defined to bee that space of time that the Sunne is in our Upper Hemisphære, to which is opposed the artificiall night, while the Sun remaineth in the lower Hemisphære. The artificiall day, as also the night, are divided each of them into 12 parts, which they call unequall houres; because that according to the different seasons of the yeare they are greater or lesse, and are never always of the same length.

The length of the artificiall day is thus found out. The Globe being set to the latitude of the place, you must find out the degree of the Eclipticke that the Sun is in at that time, and apply the same to the Meridian, and direct the Houre Index to the number of 12 in the Circle. And then turning about the Globe, till that the place of the Sun touch the Horizon at the Easterne part, the Index will Shew the houre in the Circle of the rising of the Sun; and if you but turne it about againe to the West, you shall in like manner have the houre of the setting, and so by this meanes find out the length of the artificiall day. Now if you multiply the number of the houres by 15 (for so many degrees, as we have already often said, are allowed to one equall Æquinoctiall Houre), you shall presently have the number of degrees of the Sun's Parallel that appeares above the Horizon: which if you substract out of 360, the remainder will be the quantity of that part of the same Parallel that alwaies is hid under the Horizon; or else you may proceed the contrary way, and first finde out the quantity of the Diurnall Arch, and afterward by the same you may gather the number of the houres also. For the Globe being set to the latitude of the place, and the degree of the Eclipticke that the Sunne is in beinge knowne, you may finde out, in the manner now set downe, the difference of the Right and Oblique Ascensions of the same degree of the Eclipticke for the latitude of that place. For this difference will be the halfe of that wherein the Artificiall day, for that time and place, is either deficient or exceeds the length of our Æquinoctiall day; and therefore you must adde it, when the daies are longer then the nights (which is from the 11th of March to the 12th of September), but substract all other times of the yeare, when as the nights are longer then the dayes.

Exemplum. As for example. On the 12 day of June, according to the old account, the Sunne enters into Cancer; the Right Ascension of which degree of the Eclipticke is 90 degrees. But if in the latitude of 52 gr. the first degree of Cancer bee applied to the Horizon, wee shall finde the Oblique Ascension of it to bee fiftie sixe gr. and about tenne m. So that the difference betwixt them is 33 gr. 50 min., which if you adde to ninetie gr., the halfe of the Æquinoctiall day, the length of the artificiall day will then bee 123 gr. fiftie min., and the whole Diurnall Arch 247 gr. 40 min., which if you divide by fifteene, the quotient will be sixteene and almost an halfe; which is the number of houres in the artificiall day on the twelfth of June for the latitude of fiftie two degrees.

And by this meanes may you also finde out the quantity of the longest or shortest, or any other intermediate day, together with the increase or decrease of the same, for any time or latitude of place.

Cleom. 1. Cleomedes would have the quantity of the dayes to increase and diminish after this manner; that the month immediately before, and also after the Æquinoxe, the daies should increase and decrease the fourth part of the whole difference betwixt the length of the longest and the shortest dayes of the whole yeare; and the second moneth they should differ a sixth part; and the third a twelfth part: that is if the whole difference betwixt the longest and the shortest day bee sixe houres. So that the moneth goeth immediately before, and after the Æquinoxe, the dayes increase and decrease an houre and a halfe, that is to say the fourth part of sixe houres; the second month an whole houre; and the third halfe an houre. But suppose we this to be exactly agreeable to some certaine determinate latitude, yet it is not generally so in all places. For according to the diverse Inclination of the Sphære, the daies also are observed to increase and decrease diversly. For seeing that the Parallels in every severall latitude are cut by the Æquator in a different manner, it must needs follow that the proportion of the increase and decrease of the dayes must also be different.

I shall not here need to set downe the manner how to find the apparent Arch of the Parallel of any Star, seeing that it is found out in the same manner as the Diurnall Arch of the Sunnes Parallel is.

CHAPTER XIII.

How to finde out the houre of the Day and Night, both equall and unequall, for any time or latitude of place.

If you desire to finde out the equall houre of the day, first set your Globe to the latitude of the place you are in, and also observe the latitude of the Sunne; which done, apply the place of the Sunne to the Meridian, and set the Index to the twelfth houre in the Circle, and then turne about the Globe either to the East or West, as your observation shall require, untill that the place of the Sunne be elevated so many degrees above the Horizon as shall agree with your observation, as hath been already shewed in declaring how to find the Azimuth. And the Globe standing in this situation, the Index will point in the Houre circle the houre of the day wherein your observation was made. After the same manner also you may finde the houre of the night, by observing the Altitude of any knowne Starre that is exprest in the Globe. For the Index must stand still as it did before, when it was fitted to the place of the Sunne, and the Globe must bee turned about till the Starre be observed to have the same Elevation above the Horizon of the Globe as it had in the Heavens, and then the Index will shew the houre of the night.

Now the manner how to find out the unequall houre of the day is this. First you are to find out, as we have already shewed, the quantity or number of the houres of the artificiall day, and also the equall houre of the same; whence, by the rule of proportion, you may come to the knowledge of the unequall houre.

Exemplum. In the latitude of 49 degrees the longest day containeth 16 houres. Now, therefore, when it is 10 of the clocke before Noone, or the sixth houre after Sun rising on this day, I desire to know what unequall houre of the day it is, I therefore divide my proportionall tearmes thus: 16 give 6, therefore 12 (which is the number of equall houres in every day or night) give 4 and an halfe.

And if wee desire to know how many degrees of the Æquator doe answer to one unequall houre, we may doe it thus, namely, by dividing the whole number of degrees of the Diurnall Arch by 12. As if the Artificiall day bee 16 equall houres in length, then the Arch of the Diurnall Parallel will be 240 degrees, which if we divide by 12, the quotient, which is 20, will shew the number of degrees in the Æquator that answer to one unequall houre. The like method also is to be observed in finding out the length of the unequall houre of the night.

CHAPTER XIV.

To finde out the Longitude, Latitude, and Declination of any fixed Starre as it is expressed in the Globe.

Longitudo stelle quid. The Longitude of a Starre is an Arch of Eclipticke intercepted betwixt two of the greater Circles which are drawne through the Poles of the Eclipticke, the one of which passeth through the intersection of the Æquator and Eclipticke, and the other through the Center of the Starre.

The Latitude of a Starre is the distance of it from the Eclipticke; which is also to be reckoned in that circle which passeth through the Center thereof. Latitudo quid.

Now, if you desire to find out either of these, you must take the quadrant of Altitude, or any other quadrant of a Circle that is but exactly divided into 90 parts, and lay one end of it on either Pole of the Eclipticke, either Northerne or Southerne, as the latitude of the Starre shall require. Then let it passe through the Center of the Starre to the very Eclipticke, and there the other end will shew the degree of longitude of the same, which you must reckon from the beginning of Aries, and so that portion of the Quadrant that is contained betwixt the Starre it selfe and the Eclipticke will also shew the latitude of the Starre.

The Declination of a Starre is the distance of it from the Æquator; which distance must bee reckoned on a greater circle passing through the Poles of the Æquator. And therefore if you but apply any Starre to the Meridian, you shall presently have the Declination of it, if you account the degrees and minutes of the Meridian (if there be any) that are contained betwixt the Center of the Starre and the Æquator. Declinatio quid.

CHAPTER XV.

To finde the variation of the Compasse for any Latitude of place.

That the Needle touched with the Loadstone doth decline in diverse places from the Intersection of the Meridian and Horizon is a thing most certaine, and confirmed by daily experience. Neither is this a meere forgery of Mariners, intended by them for a cloake of their own errours, as P. De

Medina, Grand Pilot to the King of Spaine, was of opinion. Neither yet doth it come to passe, by reason that the vertue of the Magnet by long use and exercise is weakened, as P. Nonius conceived, or else because it was not originally endued with sufficient vertue, as some others coldly conjecture; but this motion proceeds from its owne naturall inclination. The cause of this deflexion, although hitherto in vaine sought after by many, hath yet beene found by none. In this, as in all other of Nature's hidden and abstruse mysteries, we are quite blind. There have beene some that have endeavoured to prescribe some certaine Canon or rule for this Deflexion, as if it had beene regular and governed by some certaine order, but all in vaine. For that it is not inordinate and irregular is testified by daily experience, not only such as is taken from the dull conjecture of the common sort of Mariners, which ofttimes falls farre wide of the truth, but from the farre more accurate observations of skilful Navigatours.

At the Isles which they call Azores it declineth not at all from the true Meridian, as the common opinion of Mariners is. And I dare bee bold to affirme that at those more Western Islands also it varieth very little, or nothing at all. But if you saile Eastward from those Islands, you shall observe that point of the Needle that respects the North to incline somewhat toward the East. At Antwerp, in Brabant, it varieth about nine degrees; and neare London it declineth from the true Meridian about eleven degrees. And if you saile Westward from those Islands, the Needle also will incline toward the West. About the Sea Coasts of America, in the latitude of thirtie five or thirtie sixe degrees, it declineth above eleven degrees from the true Meridian. Beyond the Æquator it happens cleane otherwise. Neare the outwardmost Promontory of Brazile, looking Eastward, which is commonly called C. Frio, it varieth from the true Meridian above twelve degrees. Within the most Eastward parts of the Straits of Magellane it declineth five or sixe gr. And if you saile from that Promontory we now spoke of toward Africke Eastward, the variation still encreaseth, as farre as to 17 or 18 degrees, which (as farre as we can conjecture) happens in a Meridian not farre from that which passeth through the Azores. From thence the deflexion decreaseth to nine or tenne degrees, which happeneth neare the Isle of Saint Helen, bearing somewhat toward the West. And from hence they say it decreaseth till you are past the Cape of Good Hope, where they will have it to lye in the just situation of the true Meridian, neare to a certaine River, which for this cause is called by the Portugalls Rio de las Agulias. And all this deviation is toward the East.

All this wee have had certaine proofe and experience of, and that by as accurate observations as those instruments which are used in Navigation would afford, and the same examined and calculated according to the doctrine of Sphæricall Triangles. So that we have just cause to suspect the truth of many of these traditions, which are commonly delivered, concerning the deflexion of the Needle. And, namely, whereas they report that under that Meridian, which passeth through the Azores, it exactly respects the true Meridian, and that about the Sea Coasts of Brazilia the North point of the Needle declineth toward the West (as some affirme), wee have found this to bee false. And whereas they report that at New-found land it declineth toward the West above 22 degrees, we very much suspect the truth hereof, because that this seemes not at al to agree with the observation we have made concerning the variation about 11 degrees neare upon the Coast of America, of the truth of which I am so confident as of nothing more. It therefore appeares to be an idle fancy of theirs, who look to find some certaine point which the Needle should always respect; and that either on the Earth (as, namely, some certaine Magneticall Mountaines, not far distant from the Arcticke Pole), or else in the Heavens (as, namely, the taile of the little Bear, as Cardan thought), or else that it is situate in that very Meridian that passeth through the Azores, and about sixteene degrees and an halfe beyond the North Pole, as Mercator would have it. And therefore there is no need to be taken to them either, who conceive that there might be some certaine way found out of calculating the longitudes of places by means of this deflexion of the Needle, which I could wish they were able to performe; and, indeed, it might bee done, were there any certain point it should alwayes respect.

But to leave this discourse, let us now see how the quantity of this declination of the Needle may be found out by the use of the Globe, for any place of knowne latitude. And first you must provide you of some instrument by which you may observe the distance of the Suns Azimuth from the situation of a Needle. Our Mariners commonly use a Nautical Compasse, which is divided into three hundred and sixtie degrees, having a thread placed cross wise over the center of the Instrument to cast the shadowes of the Sunne upon the center of the same. This instrument is called by our Mariners the Compasse of variation; and this seemeth to be a very convenient instrument for the same use. But yet I could wish it were made with some more care and accuratenesse then Commonly it is. With this, or the like instrument, you must observe the distance of the Sunnes Azimuth, for any time or place, from the projection of the Magneticall Needle. Now we have before shewed how to find out how much the verticall circle of the Sunne is distant from the true Meridian. And the difference that there is betwixt the distance of the Sunne from the true Meridian, and from the situation of the Needle, is the variation of the Compasse. Besides, we have already shewed how the Amplitude of the rising and the setting of the Sunne may be found. If, therefore, by the helpe of this or the like instrument, it be observed (as we have said) how many degrees the Sunne riseth or setteth from those points in the Compasse that answer to the East or West, you shall in like manner have the deviation of the Needle from the true Meridian, if it have any at all.

CHAPTER XVI.

How to make a Sunne Diall by the Globe for any Latitude or Place.

We do not here promise the whole Art of Dialling; as being a matter too prolixe to be handled in this place, and not so properly concerning our present businesse in hand. And therefore it shall suffice us to have touched lightly, and, as it were, pointed out only some few grounds of this Art, being such as may very easily bee understood by the use of the Globe.

And here in this place wee shall shew you only two, the most common sorts of Dialls; one whereof is called an Horizontall Diall, because it is described on a plaine or flat which is Parallel to the Horizon; and the other is called a Murall, as being erected for the most part on a Wall perpendicular to the Horizon, and looking directly either toward the North or South. But both these may not unfitly bee called Horizontall; not in respect of the same place indeed, but of diverse. And, therefore, whether it be a Flat Horizontall, or Erect, or else Inclining any way, there will be but one kind of Artifice in making of the same.

Let us therefore now see in what manner a plaine Horizontall Diall may be made for any place. Having therefore first prepared your flat Diall ground Parallel to the Horizon, draw a Meridian on it, as exactly North and South as you possibly can. Which done, draw another East and West, which must crosse it at right angles. The first of which lines will shew twelve, and the other sixe of the Clocke, both morning and evening. Then making a Center in the Intersection of these two lines, describe a circle on your Diall ground to what distance you please, and then divide (as all other circles usually are) into 360 parts. And it will not be amisse to subdivide each of these into lesser parts, if it may conveniently be done. And now it only remaines to finde out the distances of the Houre lines in this circle for any latitude of place. Which that wee may doe by the use of the Globe, let it first be set to the latitude of the place assigned. And then make choice of some of the greater circles in the Globe, that passe through the Poles of the world (as for example the Æquinoctiall Colure, if you please): and apply the same to the Meridian, in which situation it sheweth Mid-day, or twelve of the Clocke. Then turning about the Globe toward the West (if you will), till that fifteene degrees of the Æquator have passed through the Meridian, you must marke the degree of the Horizon that the same Colure Crosseth in the Horizon. For that point will shew the distance of the first and eleventh houres from the Meridian. Both of which are distant an houres space from the Meridian or line of Mid-day. Then turning again the Globe forward, till other fifteene degrees are past the Meridian, the same Colure will point out the distance of the tenth houre, which is two houres before Noone, and of the second houre after Noone. And in the same manner you may finde out the distance of all the rest in the Horizon, allotting to each of them fifteene degrees in the Æquator crossing the Meridian. But here you must take notice by the way, that the beginning of this account of the distances must bee taken from that part of the Horizon on which the Pole is elevated; to wit, from the North part of the Horizon, if the North Pole bee elevated, and so likewise from the South part if the Antarcticke be elevated.

These distances of the Houres being thus noted in the Horizon of the Globe, you must afterward translate them into your Plaine allotted for your Diall Ground, reckoning in the circumference of it so many degrees to each houre as are answerable to those pointed out by the Colure in the Horizon. And lastly, having thus done, the Gnomon or Stile must bee erected. Where you are to observe this one thing (which is indeed in a manner the chiefe and onely thing in this Art to bee carefully looked into), namely, that that edge or line of the Gnomon, which is to show the houres by the shadow, in all kinds of Dials, must be set Parallel to the Axis of the World; that so it may make an Angle of Inclination with its plaine ground equall to that which the Axis of the World makes with the Horizon. Now that the Stile is to stand directly to the North and South, or in the Meridian line, is a thing so commonly knowne, that it were to no purpose to mention it. And this is the manner of making a Diall on a plaine Horizontall Ground.

Now if you would make a plaine Erect Diall perpendicular to the Horizon (which is commonly called a Murall), and respecting either the North or South, you must remember this one thing (the ignorance whereof hath driven those that commonly professe the Art of Dialling into many troubles and difficulties); this one thing I say is to be observed, that that which is an Erect Diall in one place will be an Horizontall in another place, whose Zenith is distant from that place 90 degrees, either Northward or Southward.

As for example: Let there be an Erect Diall made for any place whose latitude is 52 gr.[1] This is nothing else but to make an Horizontall Diall for the latitude of 38 degrees. And if there be an Erect Diall made for the latitude of 27 gr. the same will be an Horizontall Diall for the latitude of 63

[1] The 1659 edition has 25 gr.

degrees. The same proportion is to bee observed in the rest. And hence it manifestly appeares that an Horizontall Diall and a Verticall are the same at the latitudes of 45 degrees.

And so likewise by this rule may be made any manner of Inclining Diall, if so be that the quantity of the Inclination be but knowne. As, for example, if a Diall be to be made on a plaine ground, whose Inclination is 10 degrees from the Horizon Southward, and for a place whose latitude is 52 gr. Northward, you must describe in that plaine an Horizontall Diall for the latitude of 62 degrees Northward. And if in the same latitude the Diall Ground doe incline toward the North 16 gr. you must make an Horizontall Diall for the Northerne latitude of 36 gr.

And thus much shall suffice to have beene spoken of the making of Dialls by the Globe.

THE FIFTH AND LAST PART.

Of the Rombes that are described in the Terrestriall Globe, and their use.

THOSE lines which a Ship, following the direction of the Magneticall Needle, describeth on the surface of the Sea, Petrus Nonius calleth in the Latine Rumbos, borrowing the Appellation of his Countrymen the Portugals; which word, since it is now generally received by learned writers to expresse them by, we also will use the same.

These Rumbes are described in the Globe either by greater or lesser circles, or by certaine crooked winding lines. But Seamen are wont to expresse the same in their Nauticall Charts by right lines. But this practice of theirs is cleane repugnant to the truth of the thing, neither can it by any meanes be defended from errours. The invention of Rumbes, and practice of describing the same upon the Globe is somewhat ancient. Petrus Nonius hath written much concerning the use of them, in two bookes, which he intituleth *de Navigandi ratione.* And Mercator hath also expressed them in his Globes. But the use of them is not so well known to every body; and therefore I think it not unfit to be the more large in the explication of the same.

Beginning, therefore, with the nature and originall of them, we shall afterwards descend to the use there is to be made of them in the Art of Navigation. And first we will begin with the originall, and nature of the Nautical Index or Compasse; which is very well knowne to be of the fashion of a plaine rounde Boxe, the circumference whereof is divided into 32 equall parts distinguished by certaine right lines passing through the center thereof. One point of it, which that end of the Needle that is touched with the Magnet alwaies respects, is directed toward the North, so that consequently the Opposite point must necessarily respect the South. And so likewise all the other parts in it have respect unto some certaine fixed points in the Horizon (for the Compasse must alwayes be placed Parallel to the Horizon). Now I call these points fixed onely for doctrine sake, not forgetting in the meane time that the Magneticall Needle, besides that it doth of its owne nature decline in divers places from the situation of the true Meridian (which is commonly called the variation of the Compasse), according to the custome of divers Countries, is also placed after a divers manner in the Compasse. For some there are that place it 5 gr. 37 m. more Eastward then that point that answereth to the North quarter of the world, as doe the Spaniards and our Englishmen. Some place it 3 gr. and almost 18 m. declining from the North; and some set it at 11 gr. 15 m. distance from that point. All which, notwithstanding, let us suppose the Needle alwayes to look directly North and South. Now these lines thus expressed in the Mariners Compasse are the common Intersections of the Horizon and Verticall circles, or rather Parallel to these. Among which, that wherein the Needle is situate, is the common Intersection of the Horizon or Meridian. And that which crosseth this at right angles is the common section of the Horizon, and a verticall circle drawn through the Æquinoctiall East and West. And thus we have the 4 Cardinall winds or quarters of the World, and the whole Horizon divided into 4 equall parts, each of them containing 90 degrees. Now if you divide again each of these into 8 parts by 7 Verticall circles, drawne on each side of the Meridian through the Zenith, the whole Horizon will be parted into 32 equall sections, each which shall containe

11 gr. 15 m. These are the severall quarters of the world observed by Mariners in their voyages; but as for any lesser parts or divisions then these they look not after them. And this is the originall of the Nauticall Compasse by which Seamen are guided in their Voyages.

Let us now, in the next place, consider what manner of lines a Ship, following the direction of the Compasse, doth describe in her course. For the better understanding whereof I think it fit to præmise these few Propositions; which being rightly and thoroughly considered, will make the whole businesse facile and perspicuous.

1. All Meridians of all places doe passe through both the Poles, and therefore they crosse the Æquator, and all Circles Parallel to it, at right angles. Coral.

2. If wee direct our course any other way then toward one of the Poles, we change ever and anon both our Horizon and Meridian.

3. The Needle being touched with the Loadstone pointeth out the common Intersection of the Horizon and the Meridian, and one end of it alwayes respecteth the North, in a manner, and the other the South. And here I cannot but take notice of a great errour of Gemma Frisius, who, in his Corollary to the fifteene Chapter of P. Appianus Cosmography, affirmes that the Magneticall Needle respects the North Pole on this side of the Æquinoctiall line, but on the other side of the Æquinoctiall it pointeth to the South Pole. Which opinion of his is contradicted by the experience both of my selfe and others. And therefore I believe his too much credulity deceived him, giving credit perhaps to the fabulous relations of some vaine heads. But howsoever it be, the errour is a fowle one, and unworthy so great an Author. This frivolous conceit hath also beene justly condemned before by the Illustrious Jul. Scaliger, instructed hereto out of the navigations of Ludovicus Vertomannus and Ferdinand Magellane. Exer., 131, C.

4. The same Rumbe cutteth all the Meridians of all places at equall Angles, and respecteth the same quarters of the world in every Horizon.

5. A great circle drawne through the vertex of any place that is any whit distant from the Æquator cannot cut diverse Meridians at equall Angles. And therefore I cannot assent to Pet. Nonius, who would have the Rumbes to consist of portions of great circles. For, seeing that the portion of a great circle, being intercepted betwixt diverse Meridians, though never so little distant from each other, maketh unequall angles with the same, a Rumbe cannot consist of them by the præcedent proposition. But this inequality of Angles is not perceived (saith he) by the sense, unlesse it bee in Meridians somewhat farre remote from one another. Be it so. Notwithstanding, the errour of this position is discoverable by art and demonstration. Neither doth it become so great a Mathematician to examine rules of art by the judgement of the sense.

6. A great circle drawne through the Verticall point of any place, and inclining to the Meridian, maketh greater Angles with all other Meridians then it doth with that from whence it was first drawne. It therefore behoveth that a line which maketh equall angles with diverse Meridians (as the Rumbes doe) be bowed and turne in toward the Meridian. And hence it is that when a Ship saileth according to one and the same Rumbe (except it be one of the foure Principal and Cardinall Rumbes) it is a crooked and Spirall line, such as wee expressed in the Terrestriall Globe.

7. The portions of the same Rumbe, intercepted betwixt any two Parallels, whose difference of latitude is the same, are also equall to each other. Therefore an equall segment of the same Rumbe equally changeth the difference of latitude in all places. And therefore that common rule of Sea men is true: that in an equall space passed in one and the same Rumbe, one of the Poles is equally elevated and the

other depressed. So that Michael Coignet is found to be in C. 17. an errour, who, out of some certaine ill grounded positions, endeavoured to prove the contrary.

Out of the 4th Proposition there ariseth this Consectary, namely, that Rumbes, though continued never so farre, doe not passe through the Poles. For seeing that the same Rumbe is equally inclined to all Meridians—and all Meridians doe passe through the Poles—it would then follow that if a Rumbe should passe through the Poles, the same line in the same point would crosse infinite other lines; which is impossible, because that a part of any Angle cannot bee equall to the whole. Neither doth that which we delivered in the last Proposition make anything against this Consectary; to wit, that betwixt any two Parallels of equall distance, equall portions of the same Rumbe may be intercepted, that so it should thence follow that the segment of any Rumbe intercepted betwixt the Parallel of 80 gr. of latitude and the Pole is equall to a segment of the same Rumbe, intercepted betwixt the Æquator and the Parallel of tenne gr. of latitude: and the reason is, because the Pole is no Parallel. And therefore it was a true Position of Nonius that the Rumbes doe not enter the Poles, although it was not demonstrated with the like happy successe. For hee assumes foundations contrary to the truth, as wee said before. And Gemma Frisius also was mistaken when he affirmed, in his Append. ad 15 Cap. Appian, Cosmogr., that L. 2, c. 24. the Rumbes doe concurre in the Poles, which was the opinion also of some others, who are therefore justly taxed C. 17. by Michael Coignet.

These things being well considered, it will be easie to understand what manner of lines a ship, following the direction of the Magnet, doth describe in the Sea. If the forepart of the Ship be directed toward the North or South, which are the quarters that the Magneticall Needle alwayes pointeth at, your course will be alwayes under the same

Meridian: because, as wee shewed in our third Proposition, the Needle alwayes respecteth the Intersections of the Horizon and Meridian, and is situate in the plaine of the same Meridian. If the forepart of the Ship be directed to that quarter that the East and West Rumbe pointeth out, in your course you wil then describe either the Æquator or a circle Parallel to it. For if at the beginning of your setting forth your Zenith be under the Æquator, your Ship will describe an Arch or segment of the Æquator. But if your Verticall point be distant from the Æquator, either Northward or Southward, your course will then describe a Parallel, as farre distant from the Æquator as the latitude of the place is whence you set forward at first. As suppose our intended course to bee from some place lying under the Æquator, by the Rumbe of the East and West, we shall goe forward still under the Æquator. For by this meanes, as we goe on, we always meet with a new Meridian, which the line of our course crosses at right angles. Now no other line besides the Æquator can doe this; as appeares manifestly out of the Corollary of the first proposition, and therefore in this course our Ship must describe a portion of the Æquator. But if we steere our course by the East and West Rumbe from any place that lyeth besides the Æquator, we shall be alwayes under the same Parallel. For all circles parallel to the Æquator doe cut all the Meridians at right Angles, by the Corollary of the first proposition. And although the forepart of the Ship alwayes respecteth the Æquinoctiall East or West, or intersection of the Æquator and Horizon, yet in our progresse we shall never come neare the Æquator, but shall keepe alwayes an equall distance from it. Neither shall we come at all thither, whether the forepart of our Ship looketh, but shall keepe such a course, wherein we shall have ever and anon a new Meridian arising, which we shall crosse at equall Angles, and so necessarily describe a Parallel. But if our Voyage be to be made under the Rumbe which inclineth to the Meridian, our course will then be neither in a greater nor lesser circle, but we shall describe a kind of crooked spirall line. For if you draw any Greater circle through the Vertex of any place, inclining to the Meridian, the same circle will crosse the next Meridian at a greater angle than it did the former, by the 6 proposition. And therefore it cannot make any Rumbe, because the same Rumbes cutteth all Meridians at equall Angles, by the fourth proposition. And all the Parallels, or lesser circles, doe crosse the Meridians at right Angles, by the Corollary of the 1 proposition; and, therefore, they do not incline to the Meridian.

Concerning those lines which are made in sea voyages by the direction of the Compasse and Magneticall Needle. Gemma Frisius, in his Appendix to the fifteene Chapter of Appian's *Cosmography*, part 1, speaks thus: Verum hoc obiter annotandum, etc. And (saith he) I think it not amisse to note this by the way that the voyages on land doe differ very much from those that are performed at sea. For those are understood to be performed by the great circles of the Sphæres, as it is rightly demonstrated by Wernerus, in his *Commentaries* upon Ptolomy. But the voyages by sea are for the most part crooked, because they are seldome taken in a great circle, but sometimes under one of the Parallels when the Ship steers her course toward East or West, and sometimes also in a great circle, as when it saileth from North to South, or contrariwise, or else under the Æquator, either direct East or West. But in all other kinds of Navigations the journeyes are crooked, although guided by the Magnet, and are neither like to great circles, nor yet to Parallels: nor, indeed, are circles at all, but onely a kind of crooked lines, all of them at length concurring in one of the Poles. Thus hee, and, indeed, very rightly in all the rest, save onely that he will have these lines to meet in the Pole, which, as wee have already proved, is altogether repugnant to the nature of Rumbes.

Hitherto we have spoken of the originall and nature of Rumbes; let us now see what use there is of them in the Terrestriall Globe.

Of the use of Rumbes in the Terrestriall Globe.

In the Art of Navigation, which teacheth the way and manner how a Ship is to be directed in sayling from one place to another, there are some things especially to be considered. These are the longitudes of places, the latitudes, or differences of the same, the Rumbes, and the space or distance betwixt any two places, measured according to the practice used in Sea voyages. For the distances of places are measured by the Geographer one way, and by the Mariner another. For the former measureth the distance of places alwayes by great circles, as after Wernerus, Pencerus hath also demonstrated in his booke, *De Dimensione Terræ.* But the Mariners course being made up somtimes of portions of great circles, and sometimes of lesser, but for the most part of crooked lines, it is good reason that hee should measure the distances of places also by the same. Which, and how many of these are to be knowne beforehand, that the rest may be found out, comes in the next place to be considered. Now the places betwixt which our voyage is to bee performed doe differ either in longitude onely, or in latitude onely, or in both.

If they differ only in latitude they are both under the same Meridian, and therefore it is the North or South Rumbe that the course is to be directed by. And there only then remaineth to know the difference of latitude, and distance betwixt these two places: one of which being knowne, the other is easily found out. For if the difference of Latitude be given in degrees and minutes, as Sea men are wont to doe, the number of degrees and minutes being multiplyed by 60 (which is the number of English miles that we commonly allow to a degree, and that according to Ptolomies opinion, as we have already demonstrated), the whole number of miles made in the voyage betwixt these places will appeare. And if you multiply the same number of degrees by seventeene and an halfe, you have the same distance in Spanish leagues. And so contrariwise if the distance in miles or leagues be knowne, and you divide the same by 60, or seventeene and a halfe, the quotient will shew the number of degrees and minutes that answer to the differences of latitude betwixt the two places assigned. As for example. If a man were to saile from the Lizard (which is the outmost point of land in Cornewall) South ward, till he come to the Promontory of Spaine, which is called C. Ortegall, the difference of latitude of which places is 6 gr. 10 minutes; if you desire to know the distance of miles betwixt these places, multiply sixe gr. tenne m. by 60, and the product will be 370, the number of English miles betwixt the two places assigned. And this account may be much more truely and readily made by our English miles, in as much as 60 of them are equivalent to a degree, so that one mile answereth to one minute, by which means all tedious and prolixe computation by fractions is avoided.

In the next place let us consider those places that differ only in longitude, which if they lye directly under the Æquinoctiall, the distance betwixt them being knowne, the difference of longitude will also bee found, or contrariwise, by multiplication or division in like manner as the difference of latitude is found. But if they be situate without the Æquator, we must then goe another way to worke. For seeing that the Parallels are all of them lesse then the Æquator, all of them decreasing in quantity proportionably till you come to the Pole, where they are least of all; hence it comes to passe that there can be no one certaine determinate measure assigned to all the Parallels. And therefore the common sort of Mariners doe greatly erre in attributing to each degree of every Parallel an equall measure with a degree of the Æquator, by which means there have been very many errors committed in Navigation, and many whole Countryes also removed out of their owne proper situation and transferred into the places of others.

That therefore there might bee provision made in this behalfe, for those that are not so well acquainted with the Mathematiques, I have added a table, which sheweth what portion a degree in every Parallel beareth to a degree in the Æquator, whence the proper measure of every Parallel may be found. In which Table the first Colume proposeth the severall Parallels, each of them differing from other one degree of latitude. The Second sheweth the minutes and seconds in the Æquator, that answer to a degree in each Parallel; which if you convert into miles you shall know how many miles answer to a degree in every Parallel.

	M.	S.		M.	S.		M.	S.		M.	S.		M.	S.
1	59	59	27	53	27	50	38	34	71	19	31	90	0	0
2	59	57	28	52	58	51	37	46	72	18	31			
3	59	55	29	52	28	52	36	56	73	17	31			
4	59	51	30	51	57	53	36	6	74	16	31			
5	59	46	31	51	25	54	35	16	75	15	30			
6	59	40	32	50	52	55	34	24	76	14	28			
7	59	33	33	50	18	56	33	31	77	13	26			
8	59	25	34	49	44	57	32	40	78	12	24			
9	59	15	35	49	8	58	31	47	79	11	22			
10	59	5	36	48	32	59	30	53	80	10	20			
11	58	53	37	47	55	60	29	59	81	9	18			
12	58	41	38	47	17	61	29	5	82	8	16			
13	58	27	39	46	38	62	28	10	83	7	14			
14	58	13	40	45	58	63	27	14	84	6	12			
15	57	57	41	45	17	64	26	18	85	5	10			
16	57	40	42	44	35	65	25	22	86	4	8			
17	57	22	43	43	52	66	24	24	87	3	6			
18	57	3	44	43	8	67	23	26	88	2	4			
19	56	43	45	42	24	68	22	28	89	1	2			
20	56	20	46	41	40	69	21	30						
21	56	0	47	40	55	70	20	31						
22	55	37	48	40	9									
23	55	13	49	39	22									
24	54	48												
25	54	22												
26	53	55												

By the use of this Table, if a Ship have sailed under any Parallel, and the space be knowne how farre this ship hath gone, the difference of Longitude may be found by the rule of proportion ; and so contrarywise, if the difference of Longitude bee given, the distance in like manner will bee knowne. As for example; suppose a Shippe to have set forth from C. Dalguer, (which is a Promontory on the West part of Africke) and sailed Westward 200 English leagues, that is to say 600 miles. We desire now to know the difference of Longitude betwixt these two places. That Promontory hath in Northerne latitude 30 degrees, now to one degree in that Parallel answer 51 m. 57 sec., that is to say 51 miles, and fifty-seven sixtieth parts of a mile. Thus, therefore, we dispose our proportionall tearms, for the finding of the difference of Longitude 51 miles 57 min. (or suppose 52 full miles, because the difference is so small) give one degree: therefore 600 give $11\frac{28}{52}$ gr. which is the difference of Longitude betwixt the place whence the Ship set forth, and that where it arrived. But the tearmes are to be inverted if the difference of Longitude be given, and the distance be to be sought. But this is not so congruous. For we never use by the knowne Longitude to take the distance; but the contrary. Neither indeed have we as yet any certaine way of observing the difference of Longitudes; however some great boasters make us large promises of the same. But "Expectata seges vanis deludet avenis."

Cape Geer. 30° 38′ N.; de Guer of New Map; C. Dalguer of Globe, 33° 0′ N.

The hoped for crop disappointed with worthless oats.

It remaineth now to speake of those places that differ both in Longitude and Latitude; wherein there is great variety and many kinds of differences. Of all which there are foure (as we have already said) especially to be considered; and these are the differences of longitude, and of latitude, and the distance, and Rumbe by which the voiage is performed. Two of which being knowne, the rest may readily be found out. Now the transmutation of the things to be granted for knowne, and to be enquired after in these foure tearmes, may be proposed sixe manner of wayes, as followeth.

The Difference of	Longitude and Latitude	being known The	Rumbe and Distance	may also bo found.
The Difference of	Longitude and the Rumbe	being known The	Difference of Latitude and Distance	may be found.
The Difference of	Longitude and Distance	being known The	Difference of Latitude and Rumbe	may be found.
The Difference of	Latitude and Rumbe	being known The	Difference of Longitude and Distance	may be found.
The Difference of	Latitude and Distance	being known The	Rumbe and Difference of Longitude	may be found.
The	Rumbe and Distance	being known the difference of	Longitude and Latitude.	may be found.

Thus you see that any two of these being knowne, the other two may also be found out. Now most of these (yea all of them that are of any use at all) may be performed by the Globe. And let it suffice to have here given this generall advertisement once for all.

Now beside these things here already to be knowne, it is also necessary that we know the latitude of the place whence we set forth, and the quarter of the world that our course is directed unto: for otherwise we shall never be able rightly to satisfy these demands. And the reason is because that the difference of longitude and latitude is alwayes wont to be reckoned unto the two parts of the world: some of them to the North and South, and the rest to the East and West. And especially because from all parts of the Meridian, and from each side thereof, there are Rumbes drawne that are all of equall angles or inclinations. So that unlesse the quarter of the world be knowne, whereto our course tendeth, there can be no certainty at all in our conclusions. As if the difference of latitude be to be enquired after, the same may indeed be found out; but yet we cannot determine to which quarter of the world it is to be reckoned, whether North or South; and if we seeke for the difference of longitude, this may be found; but in the meane time we shall not know, whether it be to be reckoned toward the East or West. And so likewise when the Rumbe is sought for, we may perhaps find what inclination it hath to this Meridian, but yet we cannot give it its true denomination, except we know toward what quarter of the world one place is distant from the other. For from each particular part of the Meridian, the Rumbes have equall inclinations. These grounds being thus laid, let us now proceed to the examination of each particular.

I. The difference of Longitude and Latitude of two places being knowne, how to find out the Rumbe and Distance of the same.

Turne about the Globe, until that some Rumbe or other do crosse the Meridian, at the latitude of the place whence you set forth. Then again turne about either toward the East or West, as the matter shall require, untill that an equall number of degrees in the Equator to the difference of longitude of the two places do passe the Meridian. Then afterward looke whether or no the aforesaid Rumbe doe crosse the Meridian at the latitude of the place where you are, for if it does so you may then conclude that it is the Rumbe you have gone by; but if otherwise, you must take another, and try it in like manner, till you light upon one that will do it.

As for example. Serra Leona is a Promontory of Africke,
having in longitude 15 gr. 20 min., and in Northerne Lati- 13. 18 Long,
tude 7 gr. 30 m. Suppose that we are to saile to the Isle 8. 30 Lat.
of Saint Helen, which hath in longitude 24 gr. 30 m. and * 5. 41.

15. 55. N. in Southerne latitude 15 gr. 30 m., I now demand what Rumbe we are to saile by; and this we find in this manner. I first apply to the Meridian the 356 gr. 40 m. of longitude, and withall observe what Rumbe the Meridian doth crosse at the latitude Northerne of 7 gr. 30 m. (which is the latitude of the place, whence we are to set forth): and I finde it to be the North norwest, and South southeast Rumbe. Then I turne about the Globe toward the West, (because Saint Helens is more Eastward than Serra Leona untill that 9 gr. 10 m. in the Æquator, which is the difference of longitude betwixt these two places) do crosse the Meridian. And in this position of the Globe, I finde that the same Rumbe is crossed by the Meridian in the Southerne latitude of 15 gr. 30 m., which is the latitude of Saint Helens Isle. Therefore I conclude that this is the Rumbe that we are to go by, from Serra Leona to Saint Helens. And in this manner you may find the Rumbe betwixt any two places either expressed in the Globe, or otherwise; so that the difference of longitude and latitude be but knowne.

If the places be expressed in the Globe betwixt which you seeke the Rumbe; you must then with your Compasses take the distance betwixt the two places assigned, and apply the same to any Rumbe that you please (but only in those places where they crosse the Parallels of latitude of the said places) til you finde Rumbe whose portion intercepted betwixt the Parallels of the two places shal agree to the distance intercepted by the Compasses. As for example. If you would know what a Rumbe leadeth us from C. Cantin, a Promontory in the West part of Africke, having in latitude 32 gr. 20 m. to the Canary Isles, which are in the 28 gr. of latitude. First you must apply the distance intercepted betwixt the two places to any Rumbe that lyeth betwixt the 28 gr. and 32 gr. 30 m. of latitude, which are the latitudes of the places assigned: and you shall find that this distance being applyed to the South Southwest Rumbe, so that one foot of the compasses be set in the latitude of 30 gr. 20 m. the other will fall on the 28 gr. of latitude in the same Rumbe. Whence you may conclude, that you must saile from C. Cantin to the Canary Islands by the South South-west Rumbe. There are some that affirme that if this distance intercepted betwixt two places be applyed to any Rumbe where they all meet together at the Æquator the same may be performed. But these men have delivered unto us their owne errours, instead of certaine rules. For suppose it be granted that the portions of the same Rumbe intercepted betwixt two Parallels equidistant from each other, are also equall in any part of the Globe: yet notwithstanding they are not to be measured by such a manner of extension. For the Rumbes that lye neare the Æquator differ but little from greater circles, but as they are farther distant from it, so they are still more crooked and inclining to the Meridian.

The Rumbe being found, wee are next to seeke the distance betwixt the two places. Nonius teacheth a way to doe this in any Rumbe, by taking with your Compasses the space of 10 leagues, or halfe a degree. Others take 20 degrees, or an whole degree. But I approve of neither of these, nor yet regret either. Only I give this advertisement by the way, that according the greater or lesse distance from the Æquator, a greater or lesse measure may be taken. For neare the Æquator where (as we have said) the Rumbes are little different from greater circles, you may take a greater measure to goe by. But when you are farre from the Æquator you must then take as small a distance as you can, because that here the Rumbes are very crooked. And yet the distance of places may be much more accurately measured, (so that the Rumbe and difference of latitude of the same bee but knowne) by this table here set downe; which is thus:

	Rumbes.		Degr.	Min.	Sec.	
In the	First ...	...	1	1	10	Answer to a degree in the Æquator or Meridian.
	Second	...	1	4	56	
	Third ...	...	1	12	9	
	Fourth	...	1	24	51	
	Fifth ...	...	1	47	59	
	Sixth' ...	...	2	36	47	
	Seventh	...	5	7	33	

In this Table you have here set downe how many degrees, minutes, and seconds in every Rumbe do answer to a degree in the Meridian, or Æquinoctiall. Now a degree (as we have often said) containeth 60 miles; so that each mile answereth to a minute and the sixtieth part of a mile, or seventeen pases, to every second. So that by the helpe of this Table, and the rule of proportion, the distance of any two places in any Rumbe assigned (if so be that their latitude be knowne) may easily be measured; and so on the contrary if the distance be knowne, the difference of latitude may be found. As for example. If a ship have sailed from C. Verde in Africke, lying in the 14 gr. 30 m. of Northerne latitude, to C. Saint Augustine in Brazill, having in Southerne latitude, 8 gr. 30 m., by the Rumbe of Southwest and by South, and it be demanded what is the distance or space betwixt these two places. For the finding of this we dispose our tearmes of proportion after this manner, 1 gr. of latitude in this Rumbe (which is the third from the Meridian), hath 1 gr. 12 m. 9 sec., that is to say, $72\frac{9}{60}$ miles; therefore, 23 gr. (which is the difference of latitude C. Verde, and C. Saint Augustine) require 1659 miles, and almost an halfe, or something more than 553 English leagues. So that this is the distance betwixt C. Verde and C. Saint Augustine, being measured in the third Rumbe from the Meridian.

II. The Rumbe being known, and difference of Longitude; how to find the difference of latitude and distance.

To find out this you must turn the Globe till you meet with some place where the said Rumbe crosseth the Meridian at the same latitude that the place is of where you set forth. And then turning the Globe either Eastward or Westward, as you see cause, untill that so many degrees of the Æquator have passed the Meridian, as are answerable to the difference of longitude betwixt the two places; you must marke what degree in the Meridian the same Rumbe cutteth. For that degree sheweth the latitude of the place you are arived.

As for example the Isle of Saint Helen, hath in longitude 24 gr. 20 m., and in Southerne latitude 15 gr. 30 m. Suppose therefore a Shippe to have sailed West North-west, to a place that lyeth West from it 24 degrees. We demand what is the latitude of this place. First, therefore, we set the Globe in such sort, as that this Rumbe may crosse the Meridian at the 15 gr. 30 m. Southerne latitude, which is the latitude of Saint Helen, and this will happen to be so, if you apply the 37 gr. of longitude to the Meridian. Then we turne about the Globe Eastward, till that 24 gr. of the Æquator have passed under the Meridian. And then marking the degree of the Meridian, that the same Rumbe crosseth, we finde it to be about the 15 gr. 30 m. of Southerne latitude. This, therefore, we conclude to be the latitude of the place where we are arived. **Exemplum.**

And by this means also the distance may easily be found, if the Rumbe and difference of latitude be first knowne.

III. The difference of Longitude and distance being given, how to find the Rumbe and difference of Latitude.

There is not any thing in all this Art more difficult and hard to bee found than the Rumbe out of the distance and difference of Longitude given. Neither can it be done on the Globe without long and tedious practise, and many repetitions and mensuration. The practise hereof being therefore so prolixe, and requiring so much labour, it is the lesse necessary, or, indeed, rather of no use at all. And the reason is because the difference of Longitude, as wee have already shewed, is so hard to bee found out. The invention whereof I could wish our great boasters would at length performe, that so wee might expect from them something else besides bare words, vaine promises, and empty hope.

Some of these conclusions also which wee have here set downe are, I confesse, of no great use or necessity, out of the like supposition of the difference of latitude. Notwithstanding, for as much as the practise of them is easie and facile, I have willingly taken the paines, for exercise sake onely, to propose them.

IV. The difference of latitude and Rumbe being given, how to finde the difference of longitude and distance.

First set your Globe so, as that the Rumbe assigned may crosse the Meridian at the same latitude that the place is of whence you set forth, and then turne about the Globe toward the East or West, as neede shall require, till that the same Rumbe shall crosse the Meridian at the equall latitude of that place whither you have come; and so marking both places, reckon the number of degrees in the Æquator intercepted betwixt both their Meridians. And this shall be the difference of longitude betwixt the same places. As for example, C. Dalguer in Africke hath about 30 gr. of Nor- **Exemplum.** therne latitude. From whence suppose a ship to have sailed North-West and by West to the thirtie-eight gr. of Northerne latitude also. Now wee demand what is the difference of longitude betwixt these two places? Turning therefore the Globe till the Meridian crosse the said Rumbe at the thirtieth gr. of Northerne latitude (which will bee when the seventh gr. of longitude toucheth the Meridian), I turne it againe toward the East, untill such time as the Meridian crosseth the same Rumbe in the thirtie-eighth gr. of Northern latitude, which will happen when the three hundred fiftie-second gr. of longitude commeth to the Meridian. Whence we conclude that the place where the ship is arived is Westward from C. Dalguer about fifteene degrees, and the Meridian of that place passeth through the Easterne part of Saint Michaels Islands, one of the Azores. Now how the distance may be found, the Rumbe and difference of latitude being knowne, hath beene declared already in the first proposition.

V. The difference of latitude and distance being given, the Rumbe and difference of longitude may be found.

The Rumbe may easily be found out by the table which we have before set downe; but an Example will make the matter more cleare. If a ship have sailed from the most Westerne point of Africke, commonly called C. Blanco (which lyeth in the 10 gr. 30 m. of Northerne latitude) betwixt North and West, for the space of 1080 miles, and to the 20 gr. 30 m. of Northerne latitude also; and if it be demanded by what Rumbe this course was directed, for answer hereof we proceed thus: The difference of latitude is

10 gr., and the distance betwixt these places is 1080 miles, we therefore dispose our tearmes thus, 10 gr. containe 1080 miles, therefore 1 gr. containeth 108 miles, which, if we divide by 60, we shall finde in the quotient 1 gr. 48 m., which number if you seeke in the table you shall finde it answering the fifth Rumbe. Neither is the difference betwixt that number in the Table and this here of ours above one second scruple. So that we may safely pronounce that this voyage was performed by the fifth Rumbe from the Meridian, which is North west and by West. Now the Rumbe being found, and the difference of latitude knowne, you may find out the difference of longitude by the second proposition.

VI. The Rumbe and difference being given, the difference of Longitude and Latitude may also be found.

This also may easily be performed by the help of the former Table, and therefore wee will only shew an example how it is to bee done. From the Cape of Good Hope, which is the most Southernly point of Africa, and hath in Southerne latitude about 35 degrees, a ship is supposed to have sailed North North-west (which is the second Rumbe from the Meridian) above 642 miles, or if you will, let it be full 650 miles. Now we demand the difference of latitude betwixt these two places, and this is found after this manner. First, we take the degrees and minutes that answer to a degree of latitude in the second Rumbe, and turne them into miles, and then we finde the number of these to be 64 miles 65 minutes, for which let us take full 65 miles. Now, therefore, our tearmes are thus to be disposed, 65 miles answer to 1 degree of latitude, therefore 650 will be equivalent to ten degrees of latitude, which if you substract from 35 (which is the latitude of the place whence the Shippe set forth) because the course tends toward the Æquator, the remainder will be 25 gr. of Southerne latitude, which is the latitude of the place where the Ship is arived.

Now the Rumbe being knowne, and the difference of latitude also found, the difference of longitude must be found out by the second proposition.

INDEX GEOGRAPHICUS.

A.	Long.	Lat.
Aiuari ...	303 10	2 20 *a.*
Aansia ...	71 0	43 20
Abba Dalcuria ...	86 30	13 0
Abbagarima ...	70 30	8 0
Aberdeni ...	22 20	57 20
Abest	68 40	7 30
Abiami ...	57 0	7 40
Abiolho ...	93 10	5 30 *a.*
Abo	47 50	61 0
Abragama ...	156 0	32 40
el Abrigo ...	187 10	3 30 *a.*
Abriojo ...	309 30	21 30
Acapulco ...	276 0	18 0
Acartij Ins. ...	329 0	52 0
Acbul ...	154 20	49 0
Accha ...	18 30	27 40
Acchaguas ...	101 30	5 30 *a.*
Achas ...	74 30	50 40
Achbaluch ...	158 40	50 20
Achelech Magi ...	146 40	45 0
Achem ...	132 30	3 40
Acisera ...	166 0	46 30
Achilk ...	12 40	53 40
Acna ...	77 50	49 0
Azores Ins. ...	357 0	39 0
Aczu ...	123 10	49 20
Aden ...	82 0	13 50
Adia ...	50 10	25 0 *a.*
Adu ...	105 40	5 40 *a.*
Ægyptus ...	64 0	30 0
Africa op. ...	42 0	33 20
Africa Reg. ...	40 0	10 0
Agades. r. op. ...	38 20	25 30
Agisymba r. ...	24 0	7 0
Agoch ...	68 20	13 0
Agonara ...	162 20	38 0
Agragam ...	144 30	8 20 *a.*
Agro ...	68 50	6 20
Aguadd ...	173 50	7 50 *a.*
aguada segura ...	253 30	24 0
Aguada de posos.	245 20	28 0
Aham ...	14 30	24 30
Ahamara ...	14 15	25 30
Aharach ...	27 10	13 30
Almys ...	41 0	56 50
Aianirisama ...	314 30	2 40 *a.*
Aiaz	83 20	16 0
Arazza ...	72 20	39 40
Ainoedo ...	42 40	2 20
Alachi ...	70 0	20 30
Alani mon. ...	98 40	54 20
Alacranes ...	283 0	22 0
Alagoa ...	58 40	29 40 *a.*
Alaxar ...	83 40	26 0
Alar	81 30	38 40
Albaijadi ...	55 0	17 40
noua Albion ...	235 0	50 0
Albiron ...	109 30	25 30
Albasera ...	37 20	8 0
Alboram ...	25 30	35 30
Alepo ...	72 30	38 0
Alcada ...	33 0	40 30
Aldea de Arboledas	323 30	4 50
Alerandria ...	65 0	31 20
Alexandria ...	106 50	36 20
Algasin ...	16 0	29 0
Algieri ...	33 0	35 20
Algnecet ...	63 40	26 50
Alicante ...	28 40	39 0
Alicoa ...	76 40	13 20 *a.*
Alicur ...	44 20	38 30
Alima ...	108 50	31 0
Allarch ...	83 30	38 0
Alletuia ...	70 20	10 0
Almedina ...	34 0	33 40
Almeria ...	272 15	20 0
Almiria ...	26 10	37 20
Alpes Mon ...	41 30	47 30
Algnos ...	78 0	9 40 *a.*
Alsigubas ...	147 10	38 40
Aman ...	74 0	38 0
Amara ...	63 30	5 0
Amara ...	60 10	19 20 *a.*
Amostro ...	66 30	44 30
Amazon ...	45 30	12 40
Amazones Reg. ...	323 0	13 0
las Amazonas ...	312 30	12 30 *a.*
Ambastu fl. ...	167 30	50 0
Ambion Cautria ...	56 0	5 20
Ambomo ...	162 50	4 20 *a.*
Ambon ...	164 30	3 40 *a.*
Ambrun ...	35 20	44 30
Amioane ...	75 20	12 40 *a.*
Ammon ...	59 40	27 10
Ammonis ...	45 30	28 50
Amodabat ...	109 50	23 10

	Long.	Lat.
Amsterdam ...	33 0	51 30
Amu ...	143 50	35 50
Amuncla ...	47 0	29 10
Anaclon ...	112 10	41 30
Anasi ...	19 15	33 0
Anarei Montes ...	116 0	54 30
Anazaie ...	88 0	15 0 *a.*
Ancon ...	321 0	6 20
Ancona ...	63 10	1 10
Ancona ...	43 30	43 50
Anda ...	66 0	8 10 *a.*
Andemaon ...	129 30	13 0
S. Ander ...	22 10	43 10
S. Anderos ...	222 10	10 50
Andernopoli ...	58 10	44 40
S. Andra ...	170 30	12 0
S. Andre ...	22 10	56 20
Andre ...	105 40	0 50 *a.*
S. Andreas ...	62 10	61 10
Andremiri ...	60 10	43 10
Las Anegadas ...	296 0	50 0 *a.*
Angan ...	87 30	30 30
Anglesei ...	19 50	54 0
Anglia r. ...	23 0	53 0
Angolesine ...	27 0	46 0
Angolia ...	45 10	7 10 *a.*
Angos ...	69 10	15 50 *a.*
Angote r. ...	67 0	1 6
Angue ...	301 10	50 30
Angug ...	63 0	0 30
angugui ...	68 40	6 0
anna ...	80 20	29 40
S. Anna ...	318 10	27 30 *a.*
annibi reg. ...	134 10	63 0
anossa ...	29 50	12 20
anslo ...	36 30	59 20
antigua ...	320 0	16 10
antiochia ...	72 30	39 0
antiochia ...	300 50	6 40
antiosecta ...	68 30	39 30
antipara ...	74 20	25 20 *a.*
antwerpen ...	31 20	50 30
apamia ...	66 30	43 40
apola ...	93 0	21 30
aqua ...	350 30	63 0
aqualega ...	86 40	9 50 *a.*
aquilastro ...	39 40	40 0
aquile ...	42 50	46 40
ara ...	14 20	55 10
arabia felix ...	83 0	21 0
arabia deserta ...	77 0	30 0
aracam reg. ...	132 0	25 0
aracam op. ...	129 10	24 10
arami ...	64 10	17 20
arangus mons ...	46 0	8 0
arboledas ...	272 30	25 30
arcos ...	79 30	19 0 *a.*
ardagui ...	136 20	5 10 *a.*
ardanat ...	110 50	25 0
ardoc fl. ...	104 10	48 0
aredonda ...	331 20	43 0
aremogan ...	118 0	14 10
aren ...	76 10	5 10
argel ...	84 30	15 0
argin ...	11 10	10 20
arglas ...	16 30	54 20
ariaden ...	76 50	20 30
ariander ...	109 20	39 10
arica ...	300 30	20 0 *a.*
arissa ...	29 10	8 30
aries ...	32 40	44 20
arma ...	299 20	4 40
armach ...	14 50	54 0
armenia r. ...	76 0	41 0
armeta ...	57 0	17 0
armiro ...	53 10	42 30
arnaltas mons ...	35 0	11 30
aracifes ...	169 0	10 30
arsenga ...	75 30	40 40
artawiseha fl. ...	108 20	61 0
arnaga ...	137 50	32 30
arzen ...	86 20	33 30
arsila ...	21 10	35 0
asanad desertum .	20 0	18 0
asanad putens ...	18 40	21 0
ascencion ...	353 20	18 50 *a.*
La Ascension ...	15 30	8 0 *a.*
ascention ...	293 30	29 30
aschnachua ...	62 0	10 10 *a.*
ascer ...	65 70	27 0
asia reg. ...	130 6	55 0
asiot ...	67 0	26 50
asmelech ...	158 10	47 0
asmerei motes ...	137 0	50 0
asna ...	66 30	23 0
aspeza ...	52 15	44 20
aspezi montes ...	100 0	50 40
aspicia ...	164 40	47 20
assete Imaus ...	94 30	3 30 *a.*
assuan ...	69 0	22 20
Assyria reg. ...	85 0	36 0
asta ...	88 40	18 20
asta ...	66 50	15 10 *a.*
astapus fl. ...	64 0	4 0
Asuga ...	66 40	7 20
asum ...	83 10	8 30
atacama ...	303 30	32 0 *a.*
atalaia ...	283 0	20 10
atalaia ...	291 0	29 40
atalaia ...	27 50	6 20
atansa ...	159 40	7 30 *a.*
atanalo ...	298 10	1 30 *a.*
athene ...	56 10	40 0
ana ...	142 30	27 50
andegen ...	120 30	48 20
anegada ...	316 20	19 0
anero ...	17 30	41 10
angela ...	56 50	27 10

	Long.	Lat.
angesa ...	53 50	18 40 *a.*
augustin ...	293 0	29 50
auiaprari ...	317 0	5 0
avignon ...	32 40	44 40
aulona ...	51 20	41 30
auociam ...	146 50	34 0
auriata ...	66 10	9 20 *a.*
ausburg ...	38 40	48 30
ausun ...	130 30	32 10
aux ...	27 40	43 40
auzichi ...	18 40	26 30
axa ...	244 30	38 50
ayaman Reg. ...	82 20	25 0
ayque cheuonda ...	309 30	49 50
ayavire ...	306 10	18 40 *a.*
azabar ...	75 30	51 20
azamor ...	18 30	32 40.
azaphir ...	78 30	33 20
azari ...	17 15	32 10
azarit ...	68 50	28 20
azur mons ...	59 0	22 40
azzel ...	62 40	1 30 *a.*
B.		
Babelcut ...	107 50	20 10
Babel madeb ...	80 0	12 50
Babylon ...	82 20	33 0
Bachanti ...	86 0	47 0
Bachu ...	28 50	42 0
Bachnapa ...	72 0	4 0 *a.*
Bactriana Reg. ...	115 0	38 30
Badaios ...	19 40	38 30
Badalech ...	125 0	37 0
Bacca ...	303 30	1 0
Bacruchi ...	74 10	50 10
Bazar ...	52 20	21 40 *a.*
bagamiddi ...	59 30	5 0 *a.*
Bagamidri ...	61 30	6 0
Bagano ...	214 15	13 40
Baglan ...	111 50	28 20
Baglanca ...	114 30	28 0
bagosas lacus ...	77 10	50 40
Baha ...	88 40	24 30
Baham ...	176 30	11 10
Bahama ...	296 30	27 0
Baharam Ins. & Op.	87 20	27 30
B. anegada ...	319 50	40 20 *a.*
B. debaxos anegados	321 30	39 50 *a.*
buena Baia ...	190 20	4 40 *a.*
B. de los condes ...	320 20	43 0
B. de Culato ...	282 0	30 30
Baia dalagoa ...	56 10	32 10 *a.*
B. de fumos ...	240 20	36 0
Ba. sui fundo ...	318 40	41 30 *a.*
Baia de gente grande	303 0	54 0 *a.*
B. Hermosa ...	54 20	32 40 *a.*
Ba. S. Johan ...	309 40	40 30
B. S. Lunaire ...	321 30	49 20
Ba. de S. Migell ...	39 30	8 40 *a.*
B. Orsmora ...	312 30	41 0
B. de Pinos ...	233 0	40 30
Baia de placeles ...	349 30	1 50 *a.*
B. de Raphael ...	72 8	7 20 *a.*
B. de Salvad ...	344 0	20 0 *a.*
Baia de S. Sebastian	83 20	13 20 *a.*
Baiburt ...	74 40	42 30
Baicondel ...	131 20	34 40
Baida Reg. ...	126 0	65 0
Baiona ...	17 20	42 10
Baione ...	25 30	44 0
Balaghna ...	78 30	57 0
Balch ...	111 50	35 50
Balgada ...	69 30	5 0
Balsera ...	82 40	31 10
Bamberg ...	39 15	50 10
Bamplacot ...	138 40	12 40
Bancare fl. ...	48 20	4 0 *a.*
Banda ...	164 0	4 50 *a.*
Banda ...	111 20	15 30
Bandu ...	173 30	30 0
Barbacua ...	11 10	14 30
Barbada ...	320 15	19 50
la Barbada ...	192 50	1 50 *a.*
Barbado ...	322 0	13 0
Barbados ...	210 10	8 50
Barbara ...	83 10	11 0
Barca ...	62 15	11 10 *a.*
Barcena palus ...	65 20	1 0 *a.*
Bari	47 30	41 0
Baricia ...	60 40	4 40 *a.*
Barko ...	47 50	46 10
Barlingas ...	16 20	39 30
Barnagasso r. ...	70 0	13 0
Barquis ...	161 20	35 10
Base Bartit ...	326 40	70 30
S. Bartholome ...	194 30	14 0
Barua ...	59 40	3 0
Barua ...	73 50	9 50
Basel ...	37 10	47 50
Barodesertum ...	49 0	19 0 *a.*
Batachina ...	157 30	3 0 *a.*
Baticalla ...	111 30	12 40
Batimasa ...	12 30	13 0
Batnan ...	158 30	9 0
Batombar ...	157 50	5 20 *a.*
Baxos de Abreojo	350 0	18 0 *a.*
Baxos do Chapar .	148 30	15 0
Bax. de India ...	66 0	21 0 *a.*
Bax. de los Pargos	345 30	20 0 *a.*
Baxos de Patiano .	78 30	5 0 *a.*
Baxos de Villalobo	198 10	14 0
Baycis ...	140 0	35 40
Baye ...	77 50	65 0
Bazipir ...	140 40	37 20
Beciasa ...	65 0	10 30
Becolicus mõs ...	56 0	20 30

	Long.	Lat.
Beif	60 20	18 20 *a.*
Beigun	313 15	17 10
Beil	76 15	27 10
Belef	69 0	51 40
Belle Isle	334 0	52 20
Belisle	21 40	47 0
Beler	75 15	8 30
Belet	58 0	1 10 *a.*
Belis	23 40	34 20
Belloos r.	72 0	17 0
Belor desertu	125 0	44 0
Belt	52 30	50 0
Belugaras	57 10	28 30 *a.*
Benamataxa r. op.	55 0	26 0 *a.*
Benezueta	306 50	7 40
Bengala r.	126 0	26 30
Bengala op.	105 10	21 20
Benichas	136 0	3 50
Benigorai	26 0	28 30
Benigumi	25 0	30 0
Benni r. op.	41 0	7 40
Benisabeh	21 2	30 40
Benzerti	38 50	30 30
Bepyrus mos.	143 0	34 0
Bepyrus fl.	138 20	34 0
Bera	56 50	17 50 *a.*
Berdend	61 10	29 0
Berdoa r.	47 0	26 0
Berdoa op.	51 10	24 0
Bereou	87 0	24 50
Bereswa fl.	104 40	60 0
Berga	40 10	62 50
Bergen	30 30	60 50
Bermicho	52 20	30 40
Bernicho	47 20	44 50
Berwick	22 50	55 50
Besegario	11 0	10 20
Beslam	98 0	37 0
Bethle	138 50	25 40
Bexima	85 30	51 40
Biafar reg.	50 0	4 0
Biafar op.	42 30	6 10
Bialigrod	58 20	47 30
Bianza	150 0	2 50 *a.*
Bichest	32 30	32 40
Bichieri	65 30	31 20
Bicipuri	141 10	18 40
Biela	64 50	56 30
Bielo	60 20	60 10
Bigul	109 10	40 20
Bilan	100 40	41 30
Bilbao	23 30	43 0
Bileas	298 30	13 20 *a.*
Biledulgerid r.	37 0	29 0
Bilior	138 40	1 50 *a.*
Bima	151 20	8 20 *a.*
Bingiram	118 10	16 0
Bingiron	110 20	24 30
Bir	76 10	35 40
Biraen	131 40	2 0
Bisinagar op.	114 20	14 10
Bisinagar reg.	116 0	13 0
Bitonin	19 0	8 10
Blaskey	12 0	51 40
Blanes	31 10	42 0
Blanet	21 15	47 50
Bloe	5 30	67 0
Bobruesco	78 50	60 40
Bodon	52 30	45 30
Boetha	176 20	11 30
Boinare	309 40	11 20
Bolc	67 30	44 20
Bolcan	192 30	3 40 *a.*
Bolcan	164 30	27 0
Bolcanes	178 40	24 30
Bompruo Bepyrus Ptolom.	138 30	33 0
Bon	34 20	54 40
Bona	37 10	35 40
Borchi	43 30	3 20
Borgi	35 20	30 10
Borgse	40 30	64 0
Borno R. op.	48 30	17 10
Bornholm	40 50	55 30
Botwije	37 50	64 0
Bouenberg	34 20	56 30
bouincas	296 50	15 50
brandenberg	42 30	52 50
brasil	5 10	51 20
brasilia reg.	345 0	10 0 *a.*
braua	74 30	0 30
breid	12 30	67 0
brema	167 0	47 10
brema r. op.	138 20	17 30
bremen	35 10	53 20
brest	20 0	48 50
brest	331 0	53 0
breton	61 0	31 10
brius fl.	142 40	49 0
brod	43 40	49 30
brousensko	77 50	60 20
brosta	52 30	51 50
bruage	25 30	45 50
bruges	29 0	47 10
buarcos	17 30	40 10
buatili	61 30	7 30 *a.*
buda	48 0	47 20
budis	38 20	30 0
budomel	10 20	14 30
buenen	29 30	49 50
buge	71 20	21 0
bugia	34 30	35 10
buguli	15 0	9 15
bulga	88 30	54 30
bunace	68 30	5 20
buque	138 0	5 50
burdeux	26 0	45 10
burgiam	105 50	37 0

	Long.	Lat.
burien ...	18 50	50 20
burneo ...	145 40	4 50
burnesi ...	147 0	4 0
buro ...	60 40	24 0 *a.*
burro ...	160 30	3 20 *a.*
busdachsan ...	110 0	38 0
butuhar ...	154 0	7 30 *a.*
C.		
Caba ...	61 40	9 30 *a.*
Cabac ...	59 30	5 0
Cabaru ...	350 40	63 0
C. de bax de Abreojo	347 40	18 30 *a.*
C. de alinde ...	346 50	1 0 *a.*
C. de Saluise ...	324 40	51 30
C. del Ambar ...	83 30	2 0 *a.*
C. de S. Antonio...	289 15	22 50
C. de S. Antonio...	74 30	17 0 *a.*
C. de Arcas ...	44 30	16 20 *a.*
C. S. Augustin ...	162 0	6 30
C. de S. Augustino	354 0	8 30 *a.*
C. de las bals ...	44 50	18 30 *a.*
C. baxo ...	328 0	4 20
C. de las baxas ...	19 40	15 30
C. bedfort ...	320 0	65 30
C. de berica ...	284 20	7 40
C. blanco ...	273 20	25 20
C. blanco ...	281 20	10 30
C. blanco ...	330 10	1 0 *a.*
C. blanco ...	331 20	4 30
C. blanco ...	334 20	52 0
C. blanco ...	9 30	20 30
C. blanco ...	289 40	2 20 *a.*
C. blanco ...	151 0	22 40
C. bonet ...	348 30	62 40
C. brava ...	275 0	27 30
C. de breton ...	331 0	45 40
C. cameron ...	287 20	25 40
Cap cantin ...	17 0	32 10
C. de S. Catarina .	41 0	1 0 *a.*
C. Catoche ...	285 40	20 20
C. Chilan ...	96 30	41 30
C. Chili ...	297 40	35 40 *a.*
C. de collo ...	118 40	12 30
C. Comori ...	115 0	6 30
C. de Cocrita ...	45 30	21 40 *a.*
C. de correntes ...	261 30	20 20
c. de corintes ...	344 40	0 20 *a.*
C. de corrientes ...	65 40	23 40 *a.*
C. de Cro ...	31 30	42 10
C. croce ...	65 20	48 20
C. de crus ...	296 0	28 0
C. de crux ...	296 40	9 40
C. cur ...	135 20	5 40 *a.*
C. Dalguer ...	15 50	30 0
C. Demeicij ...	301 30	10 0
C. Derecho ...	71 30	11 20 *a.*
C. desierto ...	281 20	29 20

	Long.	Lat.
C. desperance ...	324 30	51 0
C. Doesmo ...	326 0	44 30
C. de S. Domingo .	315 20	46 40 *a.*
C. Drosey ...	13 0	51 10
C. Del Engano ...	158 0	19 20
C. de los Estanos .	340 50	1 0 *a.*
C. Falcahad ...	88 20	16 30
C. Falso ...	49 0	34 30 *a.*
C. Feare ...	305 10	32 30
C. Felix ...	84 30	14 10
C. del fierro ...	112 40	7 20 *a.*
C. Finis terræ ...	16 0	43 10
C. de Florida ...	293 20	25 30
C. de Folcos ...	21 20	35 50
C. Formoso ...	236 0	33 30
C. Formoso ...	28 0	5 0
C. de S. Franc ...	291 40	1 20
C. de S. Firc ...	335 30	47 50
C. Frio ...	341 15	24 0 *a.*
C. Froward ...	302 40	53 20 *a.*
C. del Gado ...	71 10	13 30 *a.*
C. de gardafui ...	86 20	12 30
C. de Gato ...	26 40	36 50
C. de Lopo Gonsuales	41 20	0 10 *a.*
C. de gratias a dios	289 20	14 20
C. de la Guija ...	290 20	5 40 *a.*
C. Guasco ...	300 10	27 10 *a.*
C. de S. Hele ...	338 30	43 0 *a.*
C. de S. Hele ...	294 0	30 40
C. de S. Helena, vel C. blanco	326 10	36 10 *a.*
C. Heregua ...	177 0	16 0
C. Henchua gregua	178 0	19 20
C. Hoa ...	163 15	3 0
C. Santiago ...	294 30	50 40 *a.*
C. de Santiago, vel de Orleans	323 30	49 10
C. de Satiago ...	309 0	37 30
C. S. Joan ...	323 30	48 30
C. S. John ...	62 30	67 30
C. del Isteo ...	42 10	4 0
C. de Isoletti ...	92 10	19 20
C. de las Islas ..	314 0	40 20
C. de Krin ...	13 0	53 40
C. Lacodera ...	311 40	9 30
C. Ledo ...	45 0	9 20 *a.*
C. de Lexus ...	318 20	41 20
C. de Lobos ...	45 20	14 50 *a.*
C. de Mabre ...	311 30	50 0
C. de Maio ...	82 50	15 50 *a.*
C. de S. Maria ...	77 30	24 0 *a.*
C. de S. Maria ...	327 20	35 10 *a.*
C. de S. Maria ...	9 40	21 40
C. Mendocino ...	23 40	42 0
C. de la Mola ...	36 50	6 30
C. dez Montes ...	293 40	48 30 *a.*
C. Morro Hermoso	301 0	11 0
C. Nasca ...	296 10	15 10 *a.*

	Long.	Lat.
C. Negro ...	44 30	17 40 *a*.
C. Neuado ...	232 20	41 0
Cabo de nombre de Jesus	308 10	53 0 *a*.
c. Ortegal ...	18 30	44 10
c. de Pales ...	28 30	38 0
c. de Palmas ...	348 10	1 20 *a*.
c. de Palmas ...	350 50	1 50 *a*.
c. de las Palmas ...	22 40	4 0
c. de quatro Palmas	34 10	6 0
c. Passaro ...	46 30	36 50
c. de S. Paulo ...	32 0	5 50
c. de Pennas ...	20 50	43 40
c. de Peseadores ...	277 40	28 0
c. del Platei ...	352 50	5 0 *a*.
c. de due pote ...	90 20	18 0
c. de Precile ...	89 0	42 30
c. primero ...	353 20	6 10 *a*.
c. primero ...	293 40	47 50 *a*.
c. primero ...	42 0	2 20 *a*.
c. de 3 puntas ...	28 30	5 20
c. de puntas ...	315 20	10 40
c. Rasalgate ...	96 20	22 20
c. Raso ...	317 40	8 0
c. de Raso ...	334 30	46 20
c. Real ...	327 10	47 50
c. de Roman ...	308 10	10 50
c. S. Roman ...	296 40	31 40
c. de S. Roque ...	76 50	25 10 *a*.
c. Roxent ...	16 30	38 50
c. Roxo ...	311 0	17 40
c. Roxo ...	11 0	12 0
c. Salida ...	74 0	26 10 *a*.
c. Spagia ...	349 40	63 40
c. bonæ spei ...	50 30	35 0 *a*.
c. de Spichel ...	17 0	38 40
c. de Spiegel ...	353 20	7 20 *a*.
c. de S. spirito ...	295 20	52 20
c. del Sp. Santo ...	161 10	13 10
c. de Stanolo ...	12 20	54 0
c. de Triburones ...	302 0	17 0
c. Tienot ...	329 40	52 30
c. de Torijga ...	11 30	18 20
c. de las vacas ...	53 0	33 40 *a*.
c. la Vela ...	305 10	11 50
c. S. Vincet ...	302 20	53 40 *a*.
c. S. Vincent ...	17 0	37 0
c. de virgin Maria .	308 0	52 10 *a*.
c. Viride ...	9 50	14 30
c. de bona Vista .	334 20	49 10
c. de vittoria ...	297 30	52 0 *a*.
c. del rosador ...	155 50	19 30
c. Walsingham ...	321 0	63 40
c. Tocheo ...	311 20	29 0
cabra ...	10 15	14 40
cabul ...	112 20	31 0
caburz ...	84 50	22 30
cacagiam ...	68 0	47 0
cacamba monte ...	79 0	19 30 *a*.
cachoberio ...	63 30	51 10
cachuchina r. op. .	140 30	20 0
caciansu ...	149 20	49 0
cacos ...	270 30	28 0
cacubay ...	27 20	23 50 *a*.
cadi ...	77 30	16 30 *a*.
cadir ...	105 30	1 20 *a*.
cachobach ...	135 0	27 30
cael ...	115 40	8 0
case ...	17 30	7 30
caffa ...	68 50	48 0
cahol ...	148 20	5 30
cahors ...	28 20	44 50
carcolam ...	114 30	8 40
caidu ...	163 40	51 30
caigra ...	60 40	4 0 *a*.
caijem ...	86 20	15 30
caim	154 15	44 0
caimana ...	192 50	0 20 *a*.
caiman grande ...	293 40	18 0
caimanes ...	294 40	17 15
cain ...	98 32	32 20
canam Sabadibe ...	167 30	3 0 *a*.
camdu reg. ...	136 0	47 0
caindu op. ...	137 30	47 40
caingu ...	147 40	40 50
cairo ...	67 30	30 0
calaian ...	148 50	34 20
calaiate ...	95 30	22 20
calaimanes ...	149 0	9 0
calamate ...	98 10	25 20
calamita ...	67 40	48 10
calam ...	110 40	61 40
calantan ...	138 30	4 0
calara ...	96 50	25 30
calatia ...	95 30	26 50
calbaca ...	136 10	35 10
calba ...	118 30	46 0
calburas mõs ...	50 0	20 0 *a*.
calco ...	269 40	23 20
caldaran ...	83 0	39 30
caldy ...	20 0	51 40
calecora ...	121 20	22 0
calecut ...	112 40	10 30
cales ...	29 10	50 40
caleture ...	118 10	15 0
calgada ...	63 0	6 10 *a*.
calhat ...	90 30	19 0
calibia r. ...	42 10	36 20
california ...	245 0	30 0
calinagam ...	119 30	19 0
caliz ...	20 50	36 10
cally ...	298 15	2 40
Calmar ..	42 30	57 40
camanar ...	300 20	16 30 *a*.
camandu ...	103 20	26 40
camareo ...	294 20	21 30
cambaba ...	150 0	8 10 *a*.

	Long.	Lat.
cambalu ...	161 10	51 46
camboa ...	19 20	8 30
camboya r. op. ...	142 20	11 40
cambriant ...	308 40	48 0
camburi ...	137 20	8 40
camenty ...	50 20	52 40
camp r. op. ...	143 40	12 0
campa ...	351 40	62 50
campar ...	134 30	0 10
campion ...	148 0	57 30
campu ...	162 20	39 40
camul ...	136 40	58 0
camultan ...	105 0	32 0
camur ...	62 20	17 10 *a.*
cana ...	68 0	25 40
canada ...	305 10	50 20
canaga rio ...	11 0	15 0
canagadi ...	290 40	33 30
canagora ...	134 30	32 40
canal del frayle ...	160 40	8 20
cananoa ...	328 40	24 50 *a.*
cananor ...	112 30	11 0
canaria ...	9 30	27 20
candahar ...	110 40	33 40
candia ...	59 30	35 30
candnigor ...	160 50	5 20
candua ...	114 10	6 30 *a.*
cane ...	25 50	53 50
canfa ...	118 30	27 0
cangre ...	67 20	42 40
canicol ...	276 15	15 0
caniem ...	99 0	62 40
caninos ...	62 30	69 10
cannaneral ...	292 50	27 10
cano ...	31 30	17 0
cant ...	104 0	46 50
canta fl ...	149 0	25 0
cantu olim Gange.	149 40	25 0
cannea ...	19 10	15 20
cannia ...	20 0	16 0
caona ...	259 40	31 0
capilan ...	130 10	14 10
capiapa ...	304 50	34 0 *a.*
capilamba ...	138 0	21 20
capis ...	42 10	31 0
capsa ...	40 0	27 10
carabach ..	115 0	34 0
carocaran ...	154 0	35 0
carocol ...	108 40	48 50
caraiam reg. ...	136 50	41 0
caraiam op. ...	139 50	41 0
carambis ...	68 20	44 50
caranganor ...	113 10	9 40
carao ...	85 40	42 40
carapetam ...	109 40	16 10
carasan ...	130 40	42 10
carcham ...	131 0	49 0
carchi ...	143 20	16 10
care desertu ...	115 0	54 0
carfur ...	85 20	11 10
carfga ...	78 40	20 40
cargirt ...	106 20	35 10
cariaco ...	314 0	9 0
cariai ...	288 20	10 20
caribana reg. ...	310 0	5 0
caribana ...	298 50	8 30
caribes ...	316 10	7 0
carcora ...	53 0	29 40
carcoran ...	153 10	61 30
carma ...	51 50	15 10 *a.*
cartagena ...	300 0	20 10
cartagena ...	28 20	38 20
cartago ...	299 30	3 10
carpart ...	76 0	38 20
carsi ...	148 10	35 50
carua ...	70 0	11 50
carut ...	91 30	22 40
casam ...	96 10	35 10
casena r. op. ...	38 20	17 10
casma ...	295 10	11 0 *a.*
cassar reg. ...	132 0	47 0
cassar ...	119 30	45 30
cassec ...	37 40	51 20
cassina fl. ...	121 40	61 0
cassor ...	106 30	1 50
castrone ...	73 50	58 0
castrum ...	165 10	60 40
castrum Portugalliæ	57 10	20 20 *a.*
catabathmus ...	58 15	31 30
catadubba ...	64 20	10 0
cataio reg. ...	150 0	53 0
cataisaset ...	115 0	35 10
catarain ...	156 15	14 10
catigam ...	128 0	22 40
catigora ...	173 50	58 0
catisclchebir ...	39 0	24 10
catnes ...	22 9	58 30
catwik ...	41 10	69 10
caubasi ...	95 20	47 0
cauas ...	308 10	17 20 *a.*
caneo desertu ...	47 0	25 0 *a.*
canit ...	155 30	7 0
canona ...	134 0	66 0
caxamalca ...	298 30	11 30 *a.*
caxines nunc Trugillo	287 10	14 20
cayneca ...	49 10	32 0 *a.*
cazar ...	86 20	56 30
cazelis ...	59 40	1 40
cazir ...	21 0	34 0
cazirmut ...	86 30	19 50
cebaco ...	288 20	13 0
cecicone ...	60 40	48 10
cedu ...	105 0	1 20 *a.*
cembuagan ...	155 20	6 30
cemeniar ...	59 6	10 0
cendergisia ...	115 30	11 40

	Long.	Lat.
cenu ...	298 40	7 20
cerabaro ...	290 0	8 50
ceraso ...	73 0	44 40
cerotigu ...	274 40	15 40
ceris ...	87 50	38 40
chaberis fl. ...	128 0	26 0
chaga ...	50 0	6 20 *a.*
chain ...	86 0	55 30
chalis ...	43 10	66 30
chalon ...	31 30	48 50
chalon ...	32 30	46 30
champaton ...	281 10	10 40
chansu fl. ...	55 10	14 0 *a.*
charangui ...	299 0	2 40 *a.*
charcas m. ...	310 0	24 30 *a.*
charcuon ...	70 0	8 40
chasehaer ...	91 30	57 30
chasteaux ...	335 0	53 0
chaul ...	109 40	17 30
chaysare ...	100 50	46 50
cheapanok ...	307 0	35 50
chela ...	173 0	37 0
chelm ...	51 30	51 0
chelonides paludes	51 30	21 30
chenchi ...	147 10	22 20
chencran ...	131 10	20 50
chendi ...	88 40	32 30
cheng ...	113 10	39 30
chepecen ...	99 10	41 40
chequeam ...	160 0	33 40
cheremandel ...	115 20	22 30
chesel fl. ...	106 10	46 50
chesimur' reg. ...	115 0	29 0
chesimur' op. ...	115 10	3 0
chesolitis ...	106 10	47 30
chesapink ...	308 0	38 0
chester ...	21 30	51 50
cheteal ...	279 40	14 40
chiagri ...	83 10	41 0
chialis ...	129 40	54 30
chialo ...	56 20	7 0 *a.*
chiamay lac ...	135 0	24 0
chiametlan ...	260 0	25 40
chianea ...	172 0	55 30
chiansu ...	147 30	27 0
chichane ...	303 50	14 0
chichester ...	26 10	51 0
chidleies cap. ...	326 40	67 30
chigi ...	28 40	11 30
chila ...	271 10	21 30
chilaban ...	117 40	7 10
chilachi ...	313 40	21 30 *a.*
chilan op. ...	96 20	41 10
chilchut ...	68 10	11 0
chileusin ...	153 50	42 20
chili reg. ...	305 0	30 0 *a.*
chili op. ...	299 0	36 30 *a.*
chilimazata ...	294 30	6 30 *a.*
chilue ...	226 20	43 20
chimines ...	302 0	11 0 *a.*
chimis ...	87 40	48 0
chincha ...	302 40	28 50 *a.*
chincheo ...	154 20	25 40
chinchitalis ...	139 20	54 30
chinsingan ...	12 40	15 0
chio ...	50 30	40 30
chiguisamba ...	305 30	17 0 *a.*
chira ...	282 20	10 40
chira ...	296 30	7 40 *a.*
chirmam r. ...	97 0	26 30
chirmam op. ...	98 30	27 30
chirman r. ...	95 0	36 0
chirmos ...	321 30	4 30 *a.*
chonel ...	78 30	26 50
choe ...	96 0	37 10
chuli ...	300 0	17 30 *a.*
chuquito ...	307 30	19 30
chur ...	37 0	47 0
cioca ...	134 10	1 30
ciangorid ...	167 0	54 0
ciaramicin ...	147 40	29 50
ciartiam op. ...	133 50	51 30
ciartiam reg. ...	136 30	51 0
cibelrian ...	80 30	19 10
cible ...	66 0	18 0 *a.*
cibuqueira ...	314 20	17 10
cignatan ...	268 40	18 50
cignateo ...	302 0	27 0
cilia ...	44 15	47 20
cincapura ...	136 40	1 20
cingui ...	156 0	42 30
cinna ...	67 0	41 20
cintaeola ...	111 20	13 50
cintagni ...	146 40	40 10
cipista ...	310 10	19 30 *a.*
cipribus ...	136 10	29 0
cirene ...	53 30	32 0
cirote ...	130 40	22 0
cirut ...	62 40	15 30 *a.*
citrochan ...	86 0	48 0
cini ...	47 10	66 30
claudia ...	318 30	41 20
cleartis palus ...	37 0	25 0
clermont ...	30 15	45 50
coale ...	65 0	21 30
coagueto ...	65 10	12 0
coar ...	132 40	23 0
cobina ...	102 50	29 0
cocas mons ...	79 0	47 30
cochia ...	20 10	12 10
cochin ...	114 0	9 40
cochinan ...	85 0	39 40
cofla ...	62 30	5 0 *a.*
cogigamri ...	118 20	52 0
coi ...	88 40	39 0
coiandu ...	119 40	43 10
cogiansa ...	157 50	43 20
coila ...	48 20	3 10 *a.*

	Long.	Lat.
colgoyne ...	68 40	69 20
colima ...	257 20	19 50
colipo ...	69 20	44 40
Collao reg. ...	310 0	16 0 *a.*
collo ...	35 10	35 30
colmogari ...	62 40	63 40
colmucho ...	117 30	6 40
coln ...	34 0	51 50
coloatan ...	269 0	25 20
colochi ...	312 0	21 20 *a.*
coloma ...	138 20	28 0
colosna ...	52 0	46 40
una coluna ...	179 40	30 0
com ...	95 20	35 40
comahagne ...	85 30	31 0
comania reg. ...	86 0	51 0
comania ...	68 0	50 0
comatay ...	131 40	22 20
combalich ...	115 20	56 40
comoro op. ...	115 10	7 10
comas ...	286 0	32 10
concritan desertum	47 0	23 0 *a.*
condu ...	116 50	36 0
congi ...	141 20	51 30
congu ...	147 20	49 10
coniga fl. ...	55 40	14 0 *a.*
congangui ...	152 40	44 0
coninxberg ...	49 10	55 30
connulaa ...	152 20	26 30
constantinopolis ...	61 20	44 40
copa ...	73 0	48 40
copaiopo ...	301 20	26 40 *a.*
copheo fl. ...	118 0	35 0
copini ...	129 10	20 0
copenhage ...	38 30	55 50
coquimbo ...	301 20	29 40 *a.*
cor ...	19 20	18 40
cora ...	85 10	19 20
corasan reg. ...	108 0	37 0
corazam ...	74 10	34 40
corck ...	15 40	51 40
corcora ...	67 50	5 0
corcoral ...	64 0	1 40
cordoba ...	316 20	33 0 *a.*
corea ...	31 20	7 20
corfu Ins. ...	22 0	39 30
choricho ...	42 30	1 0
Corinto ...	54 20	39 0
corniam ...	155 0	10 0 *a.*
Corol ...	141 40	9 10
Coromoran fl. ...	153 0	51 0
Coronades ...	295 30	45 0 *a.*
Corongo ...	302 40	14 20 *a.*
Corpo santo ...	84 10	7 30 *a.*
Corrigue ...	94 50	21 40
Corsean ...	90 50	25 0
Corsica ...	38 10	42 0
Cortad ...	56 50	30 0 *a.*
Corneo ...	290 20	32 10

	Long.	Lat.
Coruna ...	16 50	43 20
Corus ...	106 10	42 0
Corx ...	85 40	18 30
Corzali ...	32 20	35 0
Cosacan ...	89 0	37 20
Cosbas ...	77 30	40 20
Cosmai ...	90 10	46 40
Cosmaledo ...	79 50	16 50 *a.*
Cosmin fl.	135 30	20 0
Cospetir	124 0	33 0
Cossin op. ...	113 20	64 0
Cossin fl. ...	116 40	63 0
Cossir ...	69 50	24 50
Costagne ...	83 20	27 20
Costa duoyt ...	315 0	51 30
Costa poblada ...	247 30	26 50
Costa sana ...	242 20	29 20
Costnitz ...	36 15	47 50
Cotam reg. ...	130 0	51 0
Cotam op. ...	130 20	50 15
Cotam ...	145 30	14 50
Cotan ...	119 30	46 0
Cotenitz ...	88 30	59 0
Cotia ...	32 0	9 40
Conga ...	81 20	28 30
Coulam ...	114 30	7 40
Cousa ...	66 20	25 30
Cowno ...	53 10	55 0
Cozumel ...	286 30	19 0
Cracow ...	48 30	50 0
cremuch ...	81 10	44 50
crissa ...	53 20	40 0
croatamung ...	308 50	35 40
croatoan ...	308 0	34 30
croix blance ...	335 29	54 40
cuaba ..	307 30	21 30
cuama fl. ...	64 30	20 0 *a.*
cuara ...	72 40	23 40 *a.*
cubene ...	86 30	46 30
cuba ...	290 0	31 40
cuchia ...	127 20	53 10
cuchiao ...	311 40	19 20 *a.*
cuchibachoa ...	306 30	11 10
cucho ...	250 10	39 40
cudobe ...	129 30	17 40
cuerno ...	253 50	40 10
cui ...	138 10	8 20
cuitachi ...	89 15	40 50
culauropa ...	80 0	44 0
culiacan ...	256 30	27 0
culias ...	270 15	26 40
cumana ...	313 30	7 0
cumissa ...	50 20	27 30 *a.*
E. cumb. Isles ...	316 0	63 20
cumuca ...	119 40	20 20
cunasien ...	119 30	19 30
curamba ...	304 30	12 50 *a.*
curacoa ...	308 30	11 30
curate ...	109 40	20 10

	Long.	Lat.
curati ...	105 50	21 0
curch ...	90 40	32 0
curco ...	69 50	39 40
curiacuri ...	153 50	2 40 *a.*
curia muria ...	90 0	18 0
curiana ...	308 0	10 0
curiat ...	93 50	20 40
curm ...	120 50	31 0
curdem ...	117 20	35 0
cusistan reg. ...	87 0	32 0
custra ...	89 40	33 30
cuza ...	47 20	43 50
cuzco r. op. ...	297 20	13 30 *a.*
cuzco op. ...	301 40	17 40 *a.*
cwareook ...	304 0	33 40
cyprus ...	68 40	37 30
czercesi ...	64 50	51 10
czochloma ...	81 20	58 40
D.		
Dabul ...	110 0	16 40
Dacati ...	69 50	26 0
Dagaoda ...	18 20	7 0
Dager ...	56 20	22 10 *a.*
Dager port ...	48 40	59 40
Dagma ...	92 40	20 30
Dalaccia ...	77 0	14 20
Damascus ...	74 30	35 0
Dambili fl. ...	57 0	13 10 *a.*
Damiata ...	69 0	32 40
Damut ...	51 0	11 20
Damute ...	65 0	13 0
Dangala ...	66 15	17 30
Dangali r. ...	78 0	11 0
Dangara ...	53 50	10 50 *a.*
Dantzic ...	46 0	55 0
Dara r. op. ...	21 30	29 40
Dara ...	66 50	12 0
Daram ...	115 20	37 50
Darate ...	146 50	50 30
L. Darcies Island .	327 50	68 20
Darga ...	60 20	11 40
Darien ...	295 40	5 30
Darut ...	65 40	18 50
Daflon ...	63 0	48 40
Data ...	131 20	2 40
Dauagul ...	57 15	27 0 *a.*
Dauasi ...	98 50	49 40
Dauma r. op. ...	34 20	8 0
Debsan ...	52 10	13 30 *a.*
Decan ...	113 20	14 0
Dedma ...	56 30	56 30
Degme ...	60 30	22 40 *a.*
Dehebet ...	93 30	32 50
Deitam ...	142 50	20 0
Delgoy ...	74 30	67 20
La Desgraciada ...	211 20	20 0
Delli reg. ...	114 0	8 30
Delli op. ...	114 0	9 50
Dembia ...	56 0	3 0 *a.*
Dembra ...	61 10	13 40
Denia ...	29 20	39 20
Derbent ...	84 50	42 20
Derwind ...	47 50	57 30
Deseada ...	320 0	15 20
Destor ...	59 0	46 40
Deventer ...	33 25	51 50
Densen ...	31 20	31 20
Dsina ...	74 30	62 20
Dia ...	68 0	24 30
Diamuch ...	109 30	41 30
Diamuna fl. ...	131 0	36 0
Diepe ...	28 40	49 30
Diers cape ...	321 30	64 50
Digir ...	40 0	20 50
Dijon ...	32 0	47 0
Diram ...	79 30	12 10
Diu ...	108 0	20 50
Diulfar ...	87 30	16 40
Doam ...	89 40	27 20
Doara ...	81 0	8 0
Dobalia ...	63 20	19 0
Dobarea ...	69 50	15 40
Dobretan ...	332 40	43 10
Dobrowna ...	61 30	54 0
Docono ...	78 20	12 30
Dofarso ...	65 30	2 30
Doldel ...	52 30	18 0 *a.*
Domas caienhas ...	84 10	22 0 *a.*
Dominica ...	319 40	14 0
Domnes ...	50 30	51 50
Don fl. ...	75 0	53 20
Donatal ...	80 0	18 40 *a.*
Done ...	160 50	36 0
Donecz fl. ...	71 0	51 0
Donko ...	74 30	53 20
Dornate ...	137 50	7 30
Dorow ...	58 0	51 30
Dosa ...	59 0	27 10 *a.*
Dosime ...	86 0	21 0 *a.*
Dover ...	28 10	51 0
Drin ...	50 0	45 0
Drogebusch ...	64 40	55 20
Droger ...	332 0	57 30
Drongenes ...	4 30	66 30
Dronts ...	24 50	63 40
Druzech ...	59 20	54 40
Dubdu ...	25 0	32 50
Dubino ...	35 20	54 0
Dublin ...	16 40	53 10
Duda ...	67 40	13 40 *a.*
Dumaran ...	150 0	8 40
Duy ...	34 30	59 20
Duyhl ...	56 30	50 30
E.		
Ebaida ...	60 0	25 30
Ecsonen ...	30 15	58 10

	Long.	Lat.
Edenburg ...	22 0	55 50
Eillach ...	109 0	46 40
Einacen ...	73 0	11 0
Elbuchi ara ...	65 20	29 30
Elcama ...	41 20	37 30
Elisia ...	52 0	14 20 *a.*
Elisia ...	53 30	11 40 *a.*
Elgent ...	80 0	17 20
Elie ...	25 20	52 40
Eliobon ...	72 0	27 0
Elior ...	26 20	10 10
Reg. Elizabet forland	337 0	61 30
Eloacat ...	65 20	27 40
Embden ...	34 10	53 10
Emil ...	122 40	51 20
Endersockee ...	306 50	33 40
Enggi ...	55 10	24 30 *a.*
Ens ...	43 0	48 30
Ens ...	74 10	37 30
Ephesus ...	60 30	39 40
Ercoas ...	65 20	18 0
Erex ...	87 40	40 50
Ergas ...	86 0	38 0
Ergimul reg. ...	145 0	59 1
Ergimul op. ...	150 0	58 20
Erminio ...	151 50	23 40
Espainulies ...	110 40	40 50
Esser ...	66 50	13 0
Estade ...	305 10	47 40
Estahe atteradus Brettones	324 10	45 20
Estazia ...	318 10	17 10
Estrecho de Megallenes	305 0	53 20 *a.*
Euboia ...	56 10	41 0
Euphrates fl. ...	76 40	40 0
Euchor ...	93 20	36 50
Europa reg. ...	55 0	50 0
Exceter ...	22 10	51 0
Ezerim ...	77 0	42 0
Ezina ...	146 50	60 20
F.		
Fababien ...	67 30	3 20 *a.*
Falazi ...	61 20	15 30 *a.*
Falczin ...	57 20	47 0
Falsterhode ...	40 0	56 0
Famagosta ...	69 20	37 30
Famaluco ...	106 10	0 50 *a.*
Farallones ...	294 20	11 40 *a.*
Farallones ...	333 20	0 20 *a.*
Fargane ...	114 40	46 0
Farre ...	16 20	61 30
Fartache ...	86 40	16 10
C. Fartache ...	86 50	15 40
Faso ...	75 50	45 40
Fatigar ...	74 0	2 40
Fatnasa ...	38 10	30 10

	Long.	Lat.
Fayal ...	350 0	38 40
Feghig ...	25 30	31 0
Feia ...	85 20	21 10
Felicur ...	43 30	38 30
Fernando bueo ...	351 40	9 20 *a.*
Fessa r. and op. ...	21 50	32 50
Fierro ...	6 20	26 30
Finmark ...	47 0	69 30
Flamborough head	25 20	54 0
Flensborg ...	36 40	55 0
Florentia ...	41 10	43 40
Flores ...	353 40	39 20
Florida reg. ...	292 0	31 0
Focen ...	38 40	66 30
la Formanos ...	310 30	40 40
Formentera ...	31 10	38 50
Fortenentura ...	11 0	28 0
Foyl ...	15 50	55 30
Frayles ...	314 30	11 20
Francfort ...	36 30	50 0
Franca gromes ...	161 0	12 40
Frason ...	172 20	34 15
Fretum Gibraltar .	21 30	35 30
Fretum Davis ...	324 0	64 0
Frislant ...	351 30	62 0
Frobishers straits .	331 20	64 0
Fugio ...	159 40	45 10
Fugui ...	158 20	35 0
A Furious overfall	322 30	60 0
Fussum ...	161 40	37 10
Fungi ...	60 15	11 0 *a.*
G.		
Gabacha ...	80 50	39 10
Gacha ...	74 50	24 20
Gademes ...	41 10	26 30
Gaga ...	57 0	1 0
Gago reg. ...	25 0	8 30
Gaida ...	56 20	5 40 *a.*
Gainu r. ...	72 0	4 0
Galata ...	37 20	37 0
Gale ...	50 20	26 20 *a.*
Galiota ...	44 50	45 0
Galle ...	117 40	6 0
Gallila ...	52 15	16 0 *a.*
Gamba ...	64 40	17 30 *a.*
Gambra rio. ...	12 0	13 10
Gant ...	30 20	50 40
Garagoli ...	14 15	29 20
Garamantica vallis	51 30	16 0
Gargiza ...	62 40	12 0 *a.*
Garma ...	52 20	26 0 *a.*
Garnsey ...	22 29	49 40
Gaoga ...	55 0	22 0
Ganta ...	145 50	56 50
Gaza ...	70 50	33 10
Gazabele fl. ...	62 30	12 0 *a.*
Gebage ...	56 30	19 46 *a.*
Gedmec ...	362 0	61 40

	Long.	Lat.
Gelfeten ...	121 20	32 50
Gemanacota ...	118 40	6 0
Genaba ...	65 10	10 50 *a.*
Geneva ...	33 40	46 20
Gengorde ...	315 15	18 20
Genna ...	37 50	45 0
Genna ...	15 20	16 0
Geogan ...	58 10	21 0
Georgia ...	64 0	4 30 *a.*
Gerbala fl. ...	54 10	14 0 *a.*
Gerbo ...	42 0	32 0
Gerguth reg. ...	153 0	57 0
Germanareo ...	40 0	51 0
Gerseluin ...	24 30	32 20
Gesch ...	94 40	25 30
Gest reg. ...	106 30	26 0
Gest op. ...	107 30	26 30
Gesta ...	43 20	60 50
Genes ...	314 40	18 10
Ghez ...	21 0	6 30
Ghir fluvius ...	25 30	22 0
Ghir desertum ...	24 0	22 0
Giabel ...	71 20	15 40
Giamber ...	81 0	33 40
Giero ...	58 15	21 0 *a.*
Gieza ...	159 0	36 40
Gilan ...	94 0	39 20
Gilberts sound ...	326 50	67 0
Gilolo In. op. ...	161 30	1 10
Gindagu ...	157 30	48 10
Gindu ...	157 0	49 0
Ginduzi ...	138 0	25 10
Giralo ...	56 40	5 40 *a.*
Giras fl. ...	41 20	20 10
Girat ...	61 10	10 0 *a.*
Girgian ...	104 0	40 20
Goa ...	112 20	14 40
Godia ...	22 30	18 10
Goga ...	109 20	21 30
Glogau ...	43 50	51 25
Glosgon ...	29 0	57 0
Goozin rio ...	74 30	72 20
Goiame ...	57 0	14 0 *a.*
Goiasancigo ...	269 10	24 0
Gol. de S. Antonio	46 20	26 0 *a.*
Golfo de Bengala .	125 0	15 0
Gol. de Cayneca ...	49 0	32 30 *a.*
Gol. de Chalur ...	322 0	50 30
Go. Frio ...	45 30	20 0 *a.*
G. de S. Helena ...	48 40	33 30 *a.*
Golfo de la India .	44 20	3 40 *a.*
Gol. de los Negros	350 30	2 0 *a.*
Golfo de Papagaios	278 30	12 30
Gol. de Pichel ...	65 0	22 0 *a.*
Golfo del Rey ...	40 40	5 30
Golfo de todos santos	345 30	1 40 *a.*
Genera ...	7 30	26 30
Gorage r. ...	69 0	2 0
Goram ...	58 15	28 30
Gorgona ...	295 10	3 20
Gorides ...	81 20	43 0
Gotlant ...	45 20	57 30
Goto ...	75 30	46 30
Gousa ...	160 30	50 40
Gozen ...	17 10	31 30
Gozo ...	58 20	34 40
Granada ...	318 20	11 0
Granata ...	250 50	36 30
Granata ...	23 30	38 0
Græcia reg. ...	54 0	40 0
Gratiosa ...	357 30	39 30
Grenested ...	5 30	66 40
Greip ...	31 40	63 30
Grodek ...	56 30	51 30
Grodno ...	52 10	53 50
Groeningen ...	32 10	53 0
Groenland ...	0 0	75 0
Groye ...	21 0	47 20
Guachacal ...	303 10	10 50
Guachabamba ...	297 20	8 40 *a.*
Guachde ...	24 0	30 0
Guaden ...	21 20	28 30
Guaham ...	176 30	12 40
Guaian Cacus ...	147 30	45 20
Guaiaguil rel. S. Iago	294 30	2 30 *a.*
Guadalguibil ...	282 20	31 0
Gulabamba ...	294 5	0 10
Gualata ...	13 30	23 30
Guanaba ...	303 0	8 40
Guanape ...	294 50	8 10 *a.*
Guanaxas ...	284 0	15 30
Guangari r. op. ...	44 0	13 40
Guanima ...	303 0	24 20
Guadalupe ...	319 20	15 20
Guargala ...	37 30	25 50
Guber r. ...	27 0	9 0
Guber op. ...	29 20	10 40
Gubu ...	87 20	16 0
Gudan ...	48 20	8 50
Guegeue ...	22 50	14 0
Gues ...	87 40	29 10
Guenonda ...	302 40	46 10
Guerde ...	95 10	33 0
Guignam ...	178 0	16 40
Nova Guinea ...	180 0	5 0 *a.*
Guinea reg. ...	18 0	9 0
Gulye ...	33 30	50 40
Gunagona ...	67 30	6 0
Gustina ...	109 30	56 10
Guzuta ...	18 40	29 20
H.		
Haba ...	60 40	2 50 *a.*
Hacari ...	298 15	15 40 *a.*
Hagala ...	59 20	21 20 *a.*
Hales Island ...	337 50	63 0

	Long.	Lat.
Haliber ...	78 40	20 10
Halla ...	77 40	37 50
Hallicz ...	52 50	48 40
Hamacharic ...	68 10	30 30
Hamborg ...	37 10	53 20
Hammar ...	31 40	60 30
Hanguedo ...	310 30	52 0
Haroda ...	54 40	5 0 *a.*
Hartelpole ...	24 0	55 20
Harutio ...	304 0	25 30
Harwich ...	27 30	52 0
Hatoras ...	308 50	34 40
Hanana ...	292 10	20 0
Hebrides ...	15 20	58 0
Heidelberg ...	36 0	49 0
Heist ...	23 30	46 30
Heisant ...	19 30	48 40
Heit ...	79 40	22 40
Helel ...	23 50	31 40
Heprapolis ...	324 30	25 20
Hercules ...	69 20	32 10
dos Hermanos ...	182 40	25 0
Heti ...	99 50	30 0
Heylichland ...	33 50	66 0
Hibeleset ...	69 10	27 30
Hiere ...	63 20	12 40 *a.*
Hibernia ...	16 0	53 30
Hifuret ...	15 10	26 30
Hinbedesex ...	14 15	27 0
Hippodromus Ethiopiæ	12 30	17 20
Hircania reg. ...	100 0	40 0
Hispania reg. ...	25 0	40 0
Hispania noua reg.	280 0	13 30
Hispaniola ...	306 0	18 30
Hochelaga ...	300 50	44 10
Hoden ...	18 0	19 30
Hof ...	12 40	68 0
Holindal ...	36 10	61 0
Homey ...	61 30	52 50
Homi ...	169 20	37 0
Hormar ...	165 30	35 10
Honts Oort ...	48 30	59 0
Horno ...	12 10	66 10
Horo ...	178 20	21 10
Hugero ...	52 10	53 40 *a.*
Hul ...	25 20	6 40 *a.*
Humos ...	330 30	7 13 *a.*
Hunedo ...	324 0	51 30
Hungaria ...	50 0	48 0
Hurma ...	68 40	18 30
Hydaspes fl. ...	124 0	33 20
Hypasis fl. ...	124 0	33 0
I.		
Iabague ...	303 15	17 15
Iabo ...	306 10	22 10
Iaci ...	135 0	40 30

	Long.	Lat.
Iacubi fl. ...	93 0	48 0
Iadie ...	58 20	11 40
Iafuf ...	77 0	19 30
Iamaica ...	298 30	17 0
Iambut ...	72 30	26 30
Iameri ...	125 50	23 50
Ianaluiz ...	339 30	43 40
Ianathay ...	156 0	44 30
Ianco ...	98 40	45 40
Iangio ...	163 10	47 10
Iapara ...	141 20	7 40 *a.*
Iarchem op. ...	117 30	44 30
Iarchem reg. ...	117 30	44 0
Iapones ...	169 0	36 0
Iardines ...	189 30	9 30
Iarsey ...	23 0	49 20
Iastitem ...	42 50	28 0
Iatim ...	151 10	34 0
Iana maior ...	140 0	9 0 *a.*
Iana minor ...	150 0	9 0 *a.*
Iazni ...	77 30	20 30
Idita mons ...	164 0	54 40
Iepdip ...	30 0	58 40
Iericho ...	73 0	33 0
Ierom ...	100 10	55 0
Ierusalem ...	72 20	33 0
Iesd ...	94 40	32 0
Ighidi ...	32 50	25 0
Iguas ...	288 0	32 0
Iherud ...	58 20	1 0
Iliere ...	61 10	21 0 *a.*
Ilmont fl. ...	105 0	27 0
Imaus mons ...	128 0	39 0
India orientalis ...	135 0	26 0
Indion ...	105 40	38 0
Indus fl. ...	115 30	26 0
Inspurg ...	40 40	47 50
Tres Insulæ ...	169 20	2 0 *a.*
In de Aiman ...	146 30	19 0
Islas de don Alfonso de Aluares	202 0	8 0
I. de Assumptione.	324 0	52 30
I. de Atel ...	334 20	55 40
I. de Anes ...	310 30	11 20
I. de Anes ...	173 50	4 30
I. de Bastinado ...	293 30	10 30
I. de Benjaga ...	149 50	22 0
I. Blanca ...	316 50	14 40
I. Brava ...	1 20	14 20
I. del Canno ...	282 15	8 20
I. de S. Catelina...	334 10	27 30 *a.*
I. de Cedros ...	240 30	29 50
I. de S. Colunas...	178 50	30 30
Islas de Corales ...	194 40	9 50
I. deserta ...	178 0	31 0
Ilhas despera ...	335 0	46 40
I. de Enganno ...	130 40	5 40 *a.*
I. Falconum ...	142 30	68 20
I. de Fernandi ...	41 0	4 0

	Long.	Lat.
I. de Fernan Saronno	354 20	2 20 *a.*
I. del Fuego ...	2 30	14 20
I. del Fuego ...	181 30	27 40
I. Gallao ...	296 20	14 0 *a.*
I. del Galo ...	294 0	1 50
I. de don Galopes.	94 30	18 30 *a.*
I. de los Galopegos Maiores	281 10	4 0
I. de los Galopegos minores	277 30	1 0
I. de Galparico ...	204 20	14 40
I. de Garno ...	105 40	3 40 *a.*
I. de Garcea ...	162 20	2 0 *a.*
I. de Gonzalo Alveres	30 30	38 40 *a.*
I. de Gratia ...	97 20	6 30 *a.*
I. de Hombres Blancos	169 20	5 40 *a.*
I. de S. Iago ...	158 20	8 0 *a.*
I. de S. Ilefonso...	175 0	8 0
I. S. Joannis ...	312 0	18 0
I. S. Joan ...	325 30	42 30
I. S. Juan ...	164 30	6 0
I. S. Juan de Lisboa	84 20	26 40 *a.*
I. de Juan Miz ..	74 0	21 10 *a.*
I. de Juan de Mana	71 50	17 10 *a.*
I. de S. Julien ...	333 0	51 40
I. de los Ladrones.	177 20	15 0
I. de Languin ...	158 30	27 0
I. de Lima ...	295 10	22 30 *a.*
I. de Lobos ...	307 40	40 20
I. de Lobos ...	290 20	7 0 *a.*
I. de Manglares ...	288 40	12 0
I. de S. Maria ..	296 30	37 20 *a.*
I. de Martin Vaz .	10 40	21 40 *a.*
I. de los Martires .	181 20	22 50
I. de los Martires .	175 15	5 0
I. de Maio ...	4 30	14 30
I. de Micao ...	173 30	39 40
I. S. Michael ...	0 0	29 30
Islas des Minuaes .	358 0	20 20 *a.*
I. de los Nadados .	194 30	5 50
I. de Negros ...	155 30	10 30
Islas Negras ...	281 10	23 30
I. de Oiseaux ...	334 0	50 0
I. de Orleans ...	312 0	50 30
I. do ouro ...	128 40	1 20
I. do ouro ...	125 40	1 30
I. de Paiaros ...	314 0	12 40
I. de Palmos ...	163 20	6 0
I. de Paxaros ...	198 50	8 50
I. de Paxaros ...	234 20	28 0
I. de Perlas ...	293 10	7 0
I. de Pinos ...	292 20	21 30
I. del Poso ...	343 30	22 30 *a.*
I. del Principe ...	39 40	2 0
I. de Rees ...	162 0	25 20
Islas de Reyes ...	199 40	9 0

	Long.	Lat.
I. de Sal ...	4 10	16 30
I. Salomon ...	204 0	10 0 *a.*
Islas seccas ...	46 20	29 30
I. de Sembriro ...	130 0	9 50
I. de buenas sennales	161 0	9 30
I. de serranilla ...	294 40	15 30
I. Solis ...	347 40	10 30
I. S. Thomæ ...	38 0	0 0
I. de S. Thomas...	252 0	20 10
I. Tinhosa ...	146 0	18 10
I. de Tristan de Acuna	26 30	36 0 *a.*
I. Verde ...	353 50	45 30
I. de los dos Vezinos	195 30	6 30
I. de S. Vincent...	73 20	20 30 *a.*
I. de St. Vincent .	175 50	8 0
Ioam ...	135 0	7 30
Ioloso r. ...	24 30	6 0
Iontros ...	154 10	26 40 *a.*
Ioppe ...	71 20	34 0
Iorgowtz ...	76 30	57 40
Iorman ...	89 20	56 30
Iotama ...	57 30	35 30 *a.*
Ipadra ...	293 30	30 30
Iquares ...	53 0	8 10 *a.*
Isabella ...	305 20	18 50
Island ...	8 0	66 0
Istigias ...	110 30	39 40
Italia reg. ...	42 30	43 0
Ithra ...	57 30	4 50 *a.*
Juca ...	134 30	8 0
Iucatan reg. ...	283 0	18 0
I'ugor ...	138 0	7 50
I'uica ...	31 20	39 30
Iulibella ...	61 0	1 30
Iumbi ...	312 40	6 30

K.

	Long.	Lat.
Kalmuchi Tartari .	95 0	51 0
Kama fl. ...	86 0	60 0
Kanion ...	63 40	51 10
Karakithah Reg....	119 0	51 0
Karatzet ...	67 10	53 0
Kargapole ...	66 30	61 50
Kasakki Tartari...	103 0	51 0
Kenaner ...	56 10	61 30
Kiow ...	62 20	51 20
Kiro ...	46 40	64 40
Kirgessi ...	125 0	45 0
Kithias reg. ...	110 0	57 0
Kithay lacus ...	123 30	53 0
Kola ...	54 50	69 0
Kolenig ...	4 10	65 10
Kolunna ...	71 40	54 20
Kondori ...	93 20	61 50
Kosor fl. ...	96 40	49 0
Kossera ...	71 0	53 40
Konloay ...	65 20	64 10

	Long.	Lat.
L.		
Lacari ...	74 10	16 20
Laciema ...	24 50	39 30
Le lac de Goulesme	306 40	48 0
Lacus Annibus ...	131 0	60 10
Lacus Maracayba .	306 30	9 0
Lacus salsus ...	137 40	47 30
Ladena ...	50 30	41 30
Ladoga ...	62 10	61 40
Ladrios ...	155 20	14 0
Laghi ...	81 40	14 50
Lagnes ...	11 40	68 10
Lagos de los Coronades	295 0	44 0 *a.*
Laia ...	45 30	64 10
Lamon ...	70 30	1 50 *a.*
Lampesa ...	36 20	33 0
Lampurad ...	138 50	38 30
Lancerota ...	11 40	29 30
Langot ...	141 15	11 20
Langow ...	51 10	52 20
Lanos fl. ...	169 40	49 0
Lapusna ...	60 30	47 40
Laquille ...	310 20	49 0
Lar ...	91 10	29 0
Laredo ...	22 50	43 0
Larissa ...	70 0	33 0
Larta ...	53 0	46 0
Leghe ...	62 40	21 20 *a.*
Leekenes ...	29 30	58 0
Legula ...	55 0	10 10 *a.*
Lempa ...	247 10	16 50
Lempta ...	30 50	24 30
Leon ...	21 10	42 15
Leon ...	283 40	11 20
Leopolis ...	52 50	49 0
Lepin ...	98 0	58 40
Legnior maior ...	165 0	28 0
Legior min. ...	158 40	22 0
Lerida ...	28 20	41 30
Lesterpoint ...	335 0	62 0
Leuma ...	63 30	14 40 *a.*
Lezer ...	87 30	24 40
Lichi ...	145 30	23 0
Liek ...	50 20	53 50
Lima op. ...	296 40	23 30 *a.*
Limahorbaz ...	85 30	27 10
Limana ...	305 50	24 40
Limonia ...	72 10	44 20
Limosa ...	43 30	34 50
Linog ...	56 10	1 0 *a.*
Linga ...	139 50	3 30 *a.*
Liompo ...	160 20	34 40
Lion ...	32 40	45 40
Lion mons. ...	77 0	29 0
Liorne ...	40 20	43 30
Lipai ...	45 30	38 40
Lisboa ...	17 30	39 11
Lizard ...	18 30	15 10

	Long.	Lat.
Loest ...	30 40	50 0
Loffoet ...	38 10	69 0
Loghe ...	113 0	20 0
Lomfara ...	39 40	29 0
Lonbiero ...	318 20	18 40
London ...	25 50	51 40
London coast ...	326 20	72 0
Longur ...	134 20	10 50
Lop op. ...	134 20	53 0
Lop desertum ...	135 0	55 0
Lopeso ...	74 0	49 40
Loron ...	91 20	28 20
Losa ...	62 10	18 40 *a.*
Losaun ...	34 20	46 50
Loyrest ...	24 40	47 40
Loxa ...	293 30	3 50 *a.*
Lubec ...	38 30	53 50
Lucaio ...	299 0	27 30
Lucho ...	57 20	31 30
Lucka ...	42 10	52 0
Lugana ...	79 40	25 40
Luki ...	64 0	38 20
L. Lumleis inlet...	320 0	61 0
Lunæ Montes ...	60 0	16 0 *a.*
Luno ...	64 50	44 20
Lundi ...	19 30	51 0
Lutzko ...	54 0	50 20
Luso Ins. ...	156 0	17 0
Lybia Palus ...	33 0	23 30
Liocemedes Palus.	62 0	24 20
M.		
Maas ...	178 20	20 20
Maarazia ...	118 30	22 20
Maboga ...	62 40	13 30
Macara ...	32 20	30 10
Macare ...	76 20	20 50
Maceria ...	43 10	1 20 *a.*
Machian ...	160 40	0 30
Machlunaria ...	111 40	26 30
Machoenta ...	39 50	33 50
Machon ...	65 20	8 30
Macin ...	85 30	25 50
Macopa ...	132 50	1 10 *a.*
Macra ...	63 40	39 20
Macsin of Ilands .	62 30	75 30
Macyra Ins. ...	93 0	19 40
Madagascar ...	77 0	19 0
La Madalena ...	44 40	7 0
Madera ...	8 10	31 30
Madinga ...	32 50	13 0
Madura ...	146 30	6 50 *a.*
Mæatis palus ...	71 30	49 30
Magadaxo ...	78 0	5 10
Magalo ...	71 20	9 30 *a.*
Magora ...	77 50	18 40
Magurada ...	13 0	9 30
Mahag ...	64 20	4 30
Mahambana ...	54 0	32 0 *a.*

	Long.	Lat.
Maiaguaná	306 0	23 40
Maidas	2 40	46 30
Maima	47 20	10 40
Maiorica In.	39 50	33 0
Maitagasi	48 20	11 20
Maisaro	152 30	28 30
Malabrigo	178 50	26 0
Malaca r. op.	136 30	2 50
Malaga	23 50	37 20
Malati	78 0	32 40
Malana	75 0	38 20
Maldivar Insulæ	113 0	3 0
Malha	93 30	11 0 *a.*
Maliapor	118 0	13 20
Malines	279 40	13 40
Malor	82 40	10 20
Malorca op.	39 50	32 50
Malpelo	290 20	4 0
Malta	46 0	35 30
Mamora	155 40	0 40
Man	19 0	54 50
Manado	147 20	6 30
Manadu	157 50	0 30
Manaiba	77 10	22 20 *a.*
Manapata	78 10	20 50 *a.*
Manatenga r.	77 0	22 20 *a.*
Manda fl.	138 0	21 0
Mandalican	42 30	8 0 *a.*
Mandao r. op.	121 0	25 0
Mangalor	112 0	11 30
Mangesia	61 30	41 30
Mangi sive China reg.	150 0	37 0
Mangopa	131 10	3 10
Manica	62 50	23 30 *a.*
Manicongo reg.	46 40	5 0 *a.*
Manicongo op.	47 20	5 0 *a.*
Manilia	156 20	15 0
Maniolæ Ins.	140 30	2 0
Mansua	95 30	45 40
Mantra	79 50	7 0
Mapazo	307 30	7 40 *a.*
Mara	75 20	37 0
Maracapana	312 10	8 0
Marach	119 40	8 40
Maramma	56 40	9 0 *a.*
Maranga	281 30	19 30
Marannon fl.	323 0	7 0 *a.*
Marasia	146 30	26 40
Marata	262 0	32 30
Maratue	305 0	36 30
Marchant Ile	327 0	68 20
Marcoa	58 50	7 10 *a.*
Mardin	82 10	34 50
Mar de Bachu	92 0	45 0
Mare congelatum	345 0	64 0
Mar de India	120 0	10 0 *a.*
Mare maior	68 0	46 0
Ma mediter	50 0	35 0
Ma rubrum	75 0	20 0
Ma Vermejo	255 0	26 0
Ma del Zur	270 0	10 0
Maregui	134 30	13 0
Marei	52 0	26 10
Margarita	314 10	10 50
Margus fl.	111 30	52 0
Las Marias	260 0	22 0
Maril	86 30	17 30
Maricalperapo Ins.	130 40	14 0
Marigalante	320 0	14 50
Marinos	326 20	40 40
Marocco	20 0	30 30
Marseille	33 50	43 40
Martaban	134 30	17 10
Martinino	320 0	13 10
Maru	105 40	41 0
Masalig	23 30	30 20
Masaniz	96 20	47 40
Mascalat	86 40	22 20
Masia	280 40	13 20
Masta	47 10	26 40 *a.*
Masta	63 40	8 30
Mastagan	30 20	35 20
Matalotes	169 50	10 40
Matan	153 10	25 50
Matancos	296 0	25 0
Matcin	116 40	27 0
Mat flo	76 30	67 30
Matgua	89 30	18 20
Mayma	26 20	14 0
Mazacar	169 0	33 0
mazna	79 30	5 40
meaco	60 30	23 15 *a.*
meandrus mons.	152 0	31 30
meb	46 30	54 40
mecha	75 30	25 0
mechenderi	130 40	40 0
medano	295 0	31 10
los medanos	60 20	30 0 *a.*
medellian	20 50	39 0
medina celi	23 30	41 10
medina talnabi	73 0	27 20
medino	98 30	36 30
medra	45 20	11 20
medua	30 30	32 30
megiran	134 30	34 40
meidburg	39 40	52 0
meissen	41 0	51 10
mellegete	26 50	7 20
melilla	25 0	34 20
melinde r. op.	71 20	3 20 *a.*
melli reg. op.	15 40	12 0
meluing	48 0	54 50
memel	48 40	56 50
menacabo	134 50	4 40 *a.*
mendoza	305 50	32 50 *a.*
menlay	165 40	36 30
mens	35 50	50 0

	Long.	Lat.
mensa ...	59 0	4 0
mensuria ...	73 30	17 40
meraga ...	55 30	7 20
meren ...	93 20	39 40
meroe ...	68 20	16 15
mesab ...	32 0	28 40
mesat ...	101 50	36 50
meshet ...	85 30	52 50
meshite ...	67 30	25 30
mesopotamia ...	78 0	35 0
messa ...	17 0	19 30
messana ...	45 50	37 50
messet ...	91 30	31 50
messi ...	61 30	38 40
mestzora ...	75 10	55 0
mete ...	84 50	11 50
meti ...	53 50	13 40 *a.*
metlan ...	264 0	24 10
mette ..	106 40	23 50
metz ...	33 30	49 45
mezrata ...	47 40	30 40
mezu ...	133 50	35 40
miaco ...	170 30	37 0
miaos ...	159 0	2 30
mien r. ...	136 0	31 0
mien op. ...	139 30	29 50
miensko ...	56 40	54 50
miguel ...	297 0	4 0
milan ...	38 30	46 10
millo ...	57 50	36 50
mina ...	28 50	6 20
mindanao In. ...	159 0	8 0
mindanao op. ...	160 40	7 10
minden ...	35 30	52 40
mindoro ...	154 20	12 40
mingiu desertum .	100 0	31 0
minorca ...	34 30	40 0
mirocomonas ...	179 20	6 30
mirocomonas ...	302 20	21 40
La mocha ...	295 40	38 10 *a.*
mochestan ...	92 40	27 20
modon ...	53 20	37 0
modzir ...	59 50	52 0
mogar ...	57 10	24 30 *a.*
mogile fl. ...	59 30	54 0
moguer ...	20 0	37 50
mohimo ...	55 10	53 40
moi ...	86 20	25 30
moitagasi ...	58 40	17 50 *a.*
molalle ...	74 50	12 10 *a.*
moldavia reg. ...	55 0	47 0
molines ...	30 20	46 40
moltan ...	114 20	24 30
moluccæ Ins. ...	160 40	1 0
mombasa ...	72 0	4 50 *a.*
mombeza ...	79 0	8 10
momorancy ...	306 0	47 0
mompelier ...	31 30	44 10
mona ...	309 30	18 0
monasterio de la visione	73 30	12 0
monenstro ...	60 40	47 10
mongala ...	66 30	18 20 *a.*
mongul reg. ...	160 0	61 30
mongul op. ...	159 20	60 40
Los monges ...	208 30	9 40
monjes ...	307 30	11 30
monsia ...	73 0	8 10 *a.*
monsorate ...	319 10	15 40
montagala ...	106 20	43 0 *a.*
montagna	311 20	41 0
monte de bramidos	47 10	30 15 *a.*
mote especo ...	317 15	8 0
monte fragoso ...	344 0	12 0 *a.*
monte negro ...	44 40	17 0 *a.*
mount Ralegh ...	320 30	65 0
mont royal ...	301 0	45 40
mopox ...	301 10	10 0
mora ...	99 40	44 20
morea reg. ...	54 30	38 0
mosaik ...	68 50	55 0
mosambique reg. and op.	70 20	14 40 *a.*
mosconia reg. ...	80 0	59 0
moskow ...	70 30	55 40
mossa ...	84 30	35 0
mosul ...	84 0	34 50
mota ...	299 40	20 0
motil ...	160 40	0 0
motines ...	265 20	20 30
motros ...	22 20	56 50
mozend ...	24 20	34 30
moseenek ...	69 50	51 30
mubar ...	136 30	2 20
mugu ...	118 30	42 50
mullubaba ...	296 20	2 40 *a.*
multan ...	109 50	29 20
munia ...	67 0	28 0
munster ...	35 0	52 10
muron ...	76 0	55 40
mus ...	81 50	37 50
mut ...	102 50	32 40

N.

	Long.	Lat.
Nabarz ...	79 50	50 50
Nachaus ...	35 40	32 0
Naco ...	283 20	12 30
Nada ...	58 30	8 10 *a.*
Nagai tartari ...	97 0	53 30
Nagapatam ...	117 50	10 0
Nagari ...	151 30	26 40
Nagra ...	118 10	34 0
Nagnebar ...	130 30	4 50
Nagundi fl. ...	119 0	17 40
Naim ...	94 10	33 40
Naiman reg. ...	140 0	64 0
Naiman op. ...	140 0	65 10

	Long.	Lat.
Namen	31 10	50 0
Nantes	24 10	47 50
Napata	69 20	19 40
Napoli	45 0	41 0
Napoli	55 10	38 0
Napthali	73 0	34 30
Narbona	30 20	43 20
Narch	119 30	30 40
Nardenborg	47 10	67 50
Narsinga	119 0	18 0
Narua	56 10	60 0
Naseph	110 30	43 0
Nata	290 40	7 30
Natam	177 10	15 0
Natolia reg.	66 0	41 0
Nauaca	300 20	17 10
Nauiasi	277 10	14 10
Nazaret	72 40	34 10
Nebio	38 30	42 30
Neffaon	42 15	30 0
Negru	173 0	30 40
Neijna	300 30	2 20
Neli	57 50	2 20 *a.*
Nerpis	45 30	62 50
Nestra	35 0	28 10
Nestra	42 30	65 30
Neunox	57 0	64 20
Newcastle	23 10	55 20
Nicarea	59 30	39 30
Nicobar In.	130 30	16 40
Nicoia	284 30	10 40
Nicomedia	63 30	44 20
Nicopolis	56 30	45 0
Nieflot	57 50	59 50
Nil	22 10	10 30
Nilnes	98 40	58 30
Nilus fl.	67 20	32 0
Ninus	82 20	37 0
Nisa	36 10	44 0
Nisabul	102 10	38 40
Nisabul	105 0	34 30
Nischa	57 30	58 30
Nisni	79 40	56 0
Nissa	45 30	50 30
Nissa	52 20	44 30
Nito	285 10	12 0
Niues	318 40	16 20
Noe Mons	81 0	40 20
Norbate	80 0	17 10
Noion	30 0	49 20
Nombre de dios	294 30	9 20
Nomedalen	33 30	65 30
Normar	38 0	61 20
Norombega	315 40	43 40
Norwegia	35 0	62 0
Notium pr.	171 0	47 0
Noua	59 50	9 20 *a.*
Nouagradec	57 10	53 0
Nougrod	65 30	52 40
Nowgorod	62 50	60 30
Nowgorod	80 0	55 20
Nuba palus	53 0	17 20
Nubia reg.	57 0	13 0
Nubia op.	60 0	17 40
Nubia fl.	57 0	15 40
La Nublada	240 20	18 30
Nucana	138 0	9 30
Der Nues	31 0	57 30
Nurnberg	39 30	49 30

O.

	Long.	Lat.
Oby fl.	107 0	60 0
Occa fl.	77 30	55 40
Ochelasa	306 20	48 30
Odeschiria	116 0	13 20
Odia	138 30	12 0
Odnief	71 30	52 30
Oechardes fl.	134 20	58 0
Olant	43 30	57 0
Olleron	24 30	45 30
Olone	24 30	47 0
Omagua reg.	310 0	9 0 *a.*
Omba	54 10	66 50
Omedon	27 0	6 40
Omot	64 30	19 30
Onega fl.	56 40	64 0
Onegaburg	59 30	62 30
Onem	28 20	34 30
Onor	111 40	13 10
Onora des Reyes	337 40	23 40 *a.*
Onstea	79 40	59 20
Ooszee	47 0	57 0
Opakon	64 30	53 30
Opauli	21 10	6 0
Opin	80 20	5 40
Oran	29 40	35 0
Orbadari	69 0	17 30
Orcades	22 10	59 0
Orellana	343 10	3 0 *a.*
Orgabra	73 50	6 0
Oribon	59 10	48 30
Orixa r.	119 0	19 0
Orixa op.	118 40	20 40
Orleans	28 30	48 0
Ormuz Ins. & op.	91 20	27 30
Orsa	59 50	54 20
Orsa	41 20	61 30
Orpha	78 10	35 40
Ortona	44 30	42 40
Osca	27 30	42 10
Osil	49 10	50 30
Oslam	63 50	49 40
Osteco S. Miguel de Jumma	311 30	27 30 *a.*
Osties Iamaons	98 0	3 30 *a.*
Otinangiuel	68 30	43 20
Otronto	49 30	40 20
Otupe	293 50	7 0 *a.*

	Long.	Lat.
Oumare ...	80 30	6 0
Oxford ...	24 0	52 0
Oxus fl. ...	107 0	41 20
Oya r. ...	75 0	13 0
P.		
Paam ...	138 20	2 50
Paca ...	302 50	31 10 *a.*
Pacem ...	132 0	4 0
Pagam ...	177 40	18 0
Paganso ...	99 50	45 0
Paiale ...	241 50	31 20
Paita ...	290 30	5 10 *a.*
Palage ...	14 0	18 0
Palagosa ...	47 30	43 0
Palanduræ Insula.	108 0	11 0
Palatia ...	60 50	39 20
Paleacate ...	118 20	13 40
Paliace ...	55 40	32 0
Pallu ...	80 20	37 30
Palma ...	6 20	28 0
Palmar Rio ...	273 30	26 40
Palona ...	105 10	2 0 *a.*
Pamer ...	120 0	41 0
Pampalona ...	24 30	42 40
Panairuca ...	145 40	8 30 *a.*
Panama ...	394 30	8 10
Panassa ...	138 50	23 50
Pandan ...	121 50	30 10
Pantanalia ...	42 50	36 30
Panuco ...	270 10	22 20
Paquippe ...	306 0	34 40
Parasan ...	112 20	37 50
Pari ...	282 30	9 30
Paria ...	317 20	6 40
Pariban ...	136 20	7 50 *a.*
Paris ...	29 25	48 30
Parma ...	39 20	45 10
Pascar ...	59 40	1 20 *a.*
Pascherti ...	94 40	58 0
Pasir ...	105 20	24 30
Passan ...	41 50	48 40
Pasto ...	304 0	11 40 *a.*
Pastoco ...	297 50	0 0 *a.*
Patane ...	138 10	6 50
Patanis ...	99 10	25 0
Patenissi ...	109 0	20 40
Patrona ...	165 30	6 50 *a.*
Pauia ...	37 50	46 10
Pazanfu ...	136 20	31 0
Pazanfu ...	155 30	54 50
Pazer ...	134 20	3 20 *a.*
Pechora ...	66 50	67 0
Pechora castle ...	73 50	64 50
Pedir ...	131 10	4 0
Pefora ...	47 40	65 40
Pegu r. op. ...	135 0	20 10
Peim reg. ...	132 0	51 30
Peim op. ...	132 50	50 30

	Long.	Lat.
Penacote ...	119 30	18 30
Penda ...	74 10	5 20 *a.*
Pendaua ...	118 40	30 10
Perche ...	145 26	50 0
Perflaul ...	72 0	56 30
Perigo ...	323 10	43 20
Periperi ...	137 40	11 20
Pernou ...	53 30	58 40
Peru reg. ...	196 0	10 0 *a.*
Perusia ...	42 20	43 10
Pescara ...	34 30	30 10
Petallan ...	257 0	28 40
Petepoli ...	118 20	12 0
Pharacon ...	133 30	29 20
Philippinæ In. ...	158 0	15 0
Piader ...	91 30	25 0
Pico ...	356 40	38 20
Picora reg. ...	317 0	10 0 *a.*
Picora op. ...	316 40	9 30 *a.*
Las Piedras ...	296 40	4 0 *a.*
Pigmea ...	148 40	32 0
Pijusko ...	55 0	52 0
Pilingu ...	144 20	40 0
Pina ...	296 20	3 0
Pinegle ...	131 20	52 30
Pinego ...	61 10	64 30
Pinga ...	310 14	14 20 *a.*
Piramide ...	173 10	20 20
Pisa ...	40 30	43 40
Pisaena ...	302 0	24 40 *a.*
Pizan ...	73 0	51 30
Placentia ...	20 40	40 0
Las Playas ...	151 30	32 10
Plaia ...	45 20	21 0 *a.*
Plaia ...	231 50	31 20 *a.*
Plaia ...	63 30	24 30 *a.*
Plaias ...	273 20	26 0
Plaia de lagunas...	45 40	25 0 *a.*
Plata ...	315 0	19 50 *a.*
Plescow ...	59 10	59 0
Plimouth ...	21 10	50 50
Plingu ...	144 20	40 0
Ploosko ...	48 10	52 40
Plotzco ...	57 30	57 40
Pochant ...	140 0	26 30
Podenpasay ...	303 0	45 0
Podolia reg. ...	59 0	49 30
Poicters ...	26 30	47 20
Polonia reg. ...	53 0	50 0
Ponnoy ...	58 40	67 30
Pontanay ...	74 30	20 10 *a.*
Ponte viedro ...	17 20	42 40
Popaia ...	297 30	1 50
Poparopa Ins. ...	128 40	16 30
Poroguiman ...	304 30	45 0
buen Porto ...	177 30	2 0 *a.*
Puer agosto ...	298 20	53 0 *a.*
P. de Baldivia ...	296 10	36 30 *a.*
P. de Canoas ...	239 20	36 40

	Long.	Lat.
P. de Canallos ...	283 0	14 20
P. de Chili ...	300 20	31 0 *a.*
Por de la Conception	45 40	24 20 *a.*
P. Desire ...	313 0	47 40 *a.*
P. Escondo ...	157 40	17 10
P. Famine ...	302 50	53 10 *a.*
P. Fremos ...	44 0	4 0 *a.*
P. del Grado ...	42 10	3 50
Por. de Gaspar Rico	189 30	3 40 *a.*
Porto hondo ...	286 0	29 0
Po. S. Juliano ...	310 0	50 0 *a.*
P. de S. Lazaro ...	45 30	12 20 *a.*
P. de los Leonos .	318 20	42 30 *a.*
P. de S. Miguel ...	240 30	35 0
Puerto de la Misericordia	296 20	53 0 *a.*
Po. de Nauidad ...	264 0	21 30
P. de Negrillo ...	296 50	17 10
P. de Paxaro ...	157 20	16 20
P. port ...	17 30	41 10
P. de puerto ...	254 0	31 30
P. de quintero ...	300 30	32 10 *a.*
P. Real ...	21 30	36 40
Po. de los Reyes .	244 0	28 40
P. de don Rodrico	333 0	28 0 *a.*
Por. Salido ...	186 40	3 0 *a.*
P. Santo ...	10 0	32 30
P. de Sardinas ...	238 50	37 0
Po. de Juan Serrano	311 0	47 40 *a.*
P. de Velas ...	280 30	12 0
P. S. Vincente ...	337 20	23 50 *a.*
P. de Xalisco ...	260 40	24 10
Posilles ...	325 30	54 40
Posession ...	241 30	32 20
Postna ...	45 10	52 30
Potantr ...	51 40	40 50
Potiwlo ...	67 0	52 0
Potocalma ...	299 30	35 0 *a.*
Potossi ...	315 10	21 10 *a.*
Poueada ...	116 10	10 0 *a.*
poyossa ...	96 0	93 0
pracada ...	147 20	8 10 *a.*
prag ...	42 30	50 0
preflau ...	45 10	51 10
preslau ...	49 40	49 45
primsberg ...	48 30	55 10
proinay ...	75 20	71 0
pr. terræ austr. ...	13 0	42 0 *a.*
prussia reg. ...	50 0	54 0
przebors ...	48 30	50 50
ptolomais ...	66 40	29 40
puchio ...	296 0	6 50 *a.*
pulobarea ...	135 50	2 0 *a.*
pulisangar fl. ...	158 40	54 0
punto de Cayneca	48 30	32 40 *a.*
punta del Gada ...	85 50	11 0

	Long.	Lat.
pun. de la Galera .	295 40	40 0 *a.*
pun. de S. Helena	290 10	2 10 *a.*
pun. de S. Helena	325 20	37 30 *a.*
pun. de S. Lucas .	252 30	25 30
pun. de S. Maria .	60 30	29 0 *a.*
punto primero de Nauidad	58 30	31 0 *a.*
puripegam ..	120 20	21 40
puy ...	31 10	45 0
puza ...	86 30	25 0
Q.		
Quanzu ...	157 30	44 10
Quara ...	59 0	10 50 *a.*
Queda ...	135 20	6 40
Quelinsu ...	158 30	36 0
Queples ...	41 20	37 40
Queroa ...	134 0	10 10
Quesibi ...	88 20	24 30
Quiam fl. ...	139 0	54 0
Quiansu ...	144 40	42 30
Quicare Ins. ...	290 0	6 20
Quicari ...	287 30	12 0
Quilca ...	298 50	16 30 *a.*
Los Quillacinga ...	299 20	0 30
Quiloa r. op. ...	69 50	8 50 *a.*
Quinecho ...	268 50	26 0
Quinlete ...	303 40	34 40 *a.*
Quinzai ...	153 0	40 0
Quises ...	308 40	18 30
Quitaieno Ins. ...	353 40	1 40 *a.*
Quitainano ...	291 30	14 0
Quiticui ...	54 40	22 10 *a.*
Quito ...	293 10	0 10
Quiriminiao ...	144 0	6 0 *a.*
Quiuira r. ...	240 0	42 0
Quiuira op. ...	233 0	41 40
R.		
Rab ...	47 15	47 40 *a.*
Rabon ...	74 0	23 50
Ragusi ...	49 30	44 0
Raia ...	153 20	7 30 *a.*
Raige ...	79 10	28 0
Ramat ...	74 10	33 50
Rameses ...	68 30	30 30
Rane ...	352 40	62 40
Ranos Ins. ...	299 20	26 0
Raptu prom. ...	72 10	19 30 *a.*
Rarassa ...	118 30	26 0
Rasani ...	81 50	32 40
Razamuzes ...	26 20	30 20
Rast ...	91 0	39 30
Rauel ...	110 30	20 40
Rauenna ...	42 20	44 20
Rauora ...	92 0	59 0
Razer ...	88 30	28 40
Real ...	22 30	39 0
Redonda ...	193 10	4 30 *a.*

	Long.	Lat.
Regil ...	82 10	36 30
Renus ...	31 0	49 0
Rene ...	51 30	60 0
Reuen ...	114 30	45 0
Rey ...	94 40	37 10
Rezon ...	74 0	54 30
Rhobana ...	169 30	47 0
Rhezo ...	47 0	38 20
Rhodus ...	61 40	37 20
Rianrech ...	94 40	40 0
Ribadeo ...	19 20	43 20
Riffa ...	66 40	21 20
Riga ...	53 30	58 0
R. del Ancon ...	335 0	26 0 *a.*
R. de S. Andres...	178 10	2 30 *a.*
R. S. Antone ...	70 0	14 10 *a.*
R. Aoripana ...	337 10	2 0 *a.*
R. de Arboledas...	331 40	1 40
R. de S. Augustino.	350 0	15 30 *a.*
R. de S. Augustino.	183 10	2 30 *a.*
R. de S. Barbara...	326 40	34 0 *a.*
Rio de la Barca ...	321 40	5 0
R. de Barques ...	322 10	48 20
R. del Brazil ...	348 20	17 10 *a.*
R. de la Buelta ...	306 0	38 50
R. de la Buelta ...	325 20	3 40
R. de las Bueltas...	31 30	6 0
Rio de Buguli ...	15 0	9 0
R. de los Camarones	42 0	5 30 *a.*
R. del Campo ...	42 30	2 50
R. de la Canele ...	306 30	3 30 *a.*
Rio de Cano ...	298 40	33 10
R. de Carandia ...	322 10	35 0 *a.*
R. Catamanga ...	322 0	33 20 *a.*
R. de Chiriguana .	303 30	27 0 *a.*
R. de Cinaloa ...	258 30	30 0
R. de la Crux ...	308 40	49 40 *a.*
R. de Culpare ...	340 30	23 0 *a.*
R. Dangla ...	42 30	1 40
R. Doce ...	345 20	19 10
Rio Dolce ...	320 0	6 0
Rio Dulce ...	316 30	52 0
R. de S. Domingo.	353 0	7 50 *a.*
R. del Estremo ...	340 40	22 30 *a.*
R. de Flores ...	287 20	29 0
Rio de Foues ...	304 0	48 0
R. de S. Francisco.	351 40	10 0 *a.*
R. del Godo ...	34 20	6 20
R. del Ganelo ...	342 10	22 40 *a.*
R. de Gigantes ...	278 30	29 0
R. Grande ...	301 10	11 0
R. Grande ...	314 30	44 0
R. del Guato ...	284 30	29 30
R. de Gungun ...	348 30	13 10 *a.*
R. de la Hacha ...	314 15	10 40
R. de S. Hieronymo	183 40	3 0 *a.*
R. de S. Helena ...	348 40	16 30 *a.*
R. Hondo ...	290 0	52 10 *a.*

	Long.	Lat.
R. Hondo ...	318 50	41 50
Rio del Infante ...	40 40	5 30
R. de Infante ...	55 0	30 30 *a.*
R. de S. Juan ...	45 40	14 40 *a.*
R. de S. Juan nauidad	287 30	30 0
R. de laguana ...	326 30	12 0 *a.*
R. de laguna ...	55 10	32 50 *a.*
R. S. Laurens ...	318 10	53 0
R. de Lepeti ...	323 30	30 0 *a.*
Rio de Liampo ...	158 0	29 0
Rio de Limara ...	299 0	6 0 *a.*
R. de Manicongo...	48 20	10 0 *a.*
R. de Mecoretas...	320 0	31 10 *a.*
R. de Medano ...	283 50	29 30
R. de S. Mondego.	46 20	9 20 *a.*
R. de Montagnas .	319 40	42 20
R. de Naguin ...	157 30	30 30
R. Negro ...	324 40	32 10 *a.*
R. de la Notitia ...	316 0	14 10 *a.*
Rio de S. Olalla ...	301 40	7 40 *a.*
R. del oro ...	10 20	22 30
Rio de Paiamino...	298 0	5 0 *a.*
R. de Palmas ...	272 10	14 20
R. Panuco ...	271 50	22 30
Gran Rio de Parana	321 20	5 0 *a.*
R. de Pascua ...	334 50	6 30 *a.*
R. de lo Peleijo ...	321 30	43 0
Rio de Perla ...	292 30	29 0
R. de Perus ...	318 30	42 0
R. de pescadores...	331 20	3 10
R. de pescadores...	277 30	28 0
Rio peti ...	318 0	29 30 *a.*
R. de Pindado ...	157 33	18 20
R. de la Plata ...	326 30	36 0 *a.*
R. de S. polo ...	331 0	32 20 *a.*
R. de praia arrecifes	316 0	41 20
R. primero ...	327 40	45 0
Rio de los Reyes...	60 0	29 0 *a.*
R. Roque ...	351 40	4 10 *a.*
R. de Salo ...	335 40	24 30 *a.*
R. S. Saluador ...	326 10	33 0 *a.*
R. Saluador ...	17 0	7 0 *a.*
R. Santo ...	300 30	3 0 *a.*
Rio seco ...	295 30	31 30
R. Seco ...	293 50	30 30
R. de serrano ...	311 0	47 40 *a.*
R. del sp. santo ...	281 30	31 0
R. del sp. Santo...	60 0	27 30
R. de Teraiayon...	318 0	27 40 *a.*
R. de Tison ...	253 0	36 30
R. de los Topoios .	335 40	12 0 *a.*
R. de Turme ...	77 0	24 40 *a.*
R. Verde ...	321 10	5 20
R. Verde ...	289 0	33 30
R. de Vincente ...	323 40	4 40
R. Visto ...	319 40	42 20
The white riuer ...	308 10	51 20 *a.*

	Long.	Lat.
Ripon	35 30	55 20
Risan	9 30	47 0
Roan	27 40	48 50
Roca	311 0	11 10
Roca partida	248 0	19 0
Roncador	294 30	13 30
Rochelle	25 30	46 40
Rodhe	36 20	64 50
Rofain	150 30	48 40
Roma	42 30	42 0
Romana	107 40	42 10
Los Romeros	98 40	28 30
Rooswick	40 24	50 0
Ropaga	60 40	5 30 *a.*
Roguelay	314 10	50 0
Rossa	38 10	39 0
Rostone	72 10	57 0
Roswic	38 20	55 20
Ruened	58 20	19 40 *a.*
Russia	40 22	55 10
Rust	57 30	59 30
Rye	34 10	67 30
Rygalli	27 30	51 0 *a.*
S.		
Saba	317 30	17 20
Sabain	68 20	8 40
Sabarza	154 50	45 0
Sabia	60 40	23 40
Sablestan reg.	114 0	34 0
Sablon	333 50	53 50
Sabron	84 50	45 10
Sabrata	43 30	29 50
Sacabo	71 20	28 30
Sachadi	85 50	22 0
Sachi	113 10	42 20
Sachaf lacus	52 0	17 0
Sachion	135 50	56 30
Sacolche	68 0	15 10
Sacole	69 30	19 0
Saendebar	174 40	35 50
Saepam	176 30	14 0
Sagatin	95 30	58 20
Sala	20 20	33 30
Sala	89 40	48 0
Salamanca	20 30	40 50
Salane	63 10	13 40
Salasta	72 40	41 50
Salata	76 0	24 30
Salebrena	24 50	37 30
Salina	45 0	38 30
Salinus de Trenot.	321 40	53 0
Salsburg	42 0	48 20
Salsipodes	288 40	15 40
Salstom	32 20	62 0
Saluado	321 20	5 0
Samain	111 0	46 30
Samma	51 0	65 0
Samara	118 30	8 10
Samarchan	130 20	47 40
Samarchant	109 0	44 0
Samaria	72 20	33 40
Samirent	90 50	35 30
Samot	51 20	28 0 *a.*
Sana	84 40	17 50
Sana	70 30	23 40
Sanbicasas	78 0	12 40 *a.*
Sandace	69 20	18 0
Sandersons tower.	320 0	65 30
Hope Sanderson	326 20	72 40
Sandri	162 50	53 0
Saguenay fl.	306 40	55 0
Sanguin	160 20	41 20
Sanostol	350 0	62 0
Sanson	20 40	43 20
Santari	73 40	17 0
Sante	294 40	9 30 *a.*
S. Apolonia	82 30	21 40 *a.*
S. Barnardo	328 40	12 30
S. Barnardo	181 20	23 20
S. Barnardo	319 50	17 0
S. Bartolomeo	319 10	17 50
S. Catarina In.	308 20	17 0
S. Catharina	292 50	12 10
S. Christophero	318 30	16 40
S. Christoual	291 20	22 20
S. Christoual	306 30	38 0
S. Croce	4 0	1 0 *a.*
S. Croce	3 50	1 0 *a.*
S. Crux	314 15	16 0
S. Crux	334 20	43 30
S. Cruxde la Sierra	318 10	17 20 *a.*
S. Dauids	20 0	52 0
S. Domingo	307 10	17 50
S. Espirito	322 30	31 20 *a.*
S. Espirito	75 40	13 50 *a.*
S. Francisco	87 0	7 0 *a.*
S. Francisco	326 20	24 40 *a.*
S. Francisco	255 0	31 50
S. Francisco	335 20	26 10 *a.*
S. George	357 10	39 0 *a.*
S. Helena	24 30	16 0 *a.*
S. Hieronymo	181 30	24 0
S. Hieroms Riuer.	302 10	53 10 *a.*
Santiago	264 30	20 30
Santiago	298 10	32 10
S. Jago	175 30	2 0 *a.*
S. antiago	320 0	14 30
S. Juan	171 30	6 20
S. Jan de Luz	25 10	43 20
S. Juan de Lua	273 20	19 40
S. Lazaro	71 0	10 20 *a.*
S. Lucar	21 20	37 10
S. Lucia	1 0	17 0
S. Lucia	319 50	12 40
S. Malo	24 20	48 50
S. Maria	82 30	17 0 *a.*
S. Maria	240 40	34 20

	Long.	Lat.
S. Maria ...	0 30	36 0
S. Maria ...	356 0	18 30 *a.*
S. Maria ...	85 0	44 30
S. Maria de Nazaret	66 30	16 30
S. Martha ...	301 20	10 40
S. Martin ...	321 10	51 0
S. Martin In. ...	293 40	46 50 *a.*
S. Martin ...	319 10	17 10
S. Matheo ...	21 10	1 50 *a.*
S. Michel ...	60 50	65 30
S. Miguel ...	327 20	47 20
S. Miguel ...	291 40	6 10 *a.*
S. Miguel ...	268 0	24 0
S. Miguel ...	249 0	32 50
S. Nicolas ...	69 0	64 0
S. Nicholas ...	323 20	53 40
S. Nicholas ...	2 0	17 0
S. Pietro ...	64 30	1 30
S. Pol ...	330 40	47 20
S. Pol. de Lyon ...	20 40	48 40
S. Samson ...	306 30	40 30
S. Vincent ...	0 30	17 30
S. Vincente ...	318 40	41 50
S. Vues ...	17 30	38 30
Todos Santos ...	319 10	14 50
Todos Santos ...	350 30	12 30 *a.*
Sanx Pouan ...	75 30	14 40 *a.*
La Saona ...	309 0	16 50
Sapom Ins. ...	107 10	0 30
Sarachi ...	84 30	44 10
Saragosa ...	26 10	41 50
Saragna ...	64 0	9 0
Sararatzik ...	87 10	48 30
Sardinia ...	39 0	40 0
Sargora ...	54 30	15 20 *a.*
Sargora ...	137 40	40 0
Sari ...	97 30	37 40
Sartan ...	84 0	14 40
Sata ...	72 50	25 40
Satyroru Ins. ...	174 10	46 30
Sana ...	81 10	18 40
Santopoli ...	73 30	47 0
Saun ...	115 20	47 20
Saura ...	84 40	31 20
Scampi ...	51 50	42 0
Scanga ...	119 30	26 0
Scarborough ...	24 50	54 30
Scarpanto ...	62 10	36 0
Scierno ...	145 15	28 0
Sciro ...	57 30	41 10
Schotland ...	25 0	60 0
Schwitenes ...	28 50	59 0
Scosna ...	75 40	52 0
Scotia r. ...	20 0	57 0
Scudo ...	291 0	9 0
Scylazo ...	47 40	39 20
Secotan ...	304 50	34 30
Segedein ...	49 0	47 10
segn nuestra Ins. .	293 30	46 20 *a.*
selefer ...	135 50	33 20
selg ...	111 50	48 0
semes ...	19 30	48 20
semon ...	95 40	38 10
senega reg. ...	13 0	24 0
senega fl. ...	12 30	11 30
septa ...	22 0	35 40
seretus ...	118 50	28 50
sereng ...	105 30	27 50
serent ...	97 10	29 0
serneri ...	103 30	35 0
serneri reg. ...	106 33	33 30
seroponon ...	115 40	59 20
serra ...	158 20	49 10
serra liona ...	15 30	7 40
serras de S. Esperitu	42 30	1 50 *a.*
serta ...	62 0	26 50
seruan ...	90 20	39 45
seu desertum ...	52 0	14 30
shaboglishar ...	83 40	56 30
shakaskik ...	91 30	53 0
shensk ...	68 40	61 50
skalholt ...	8 30	65 20
sian ...	139 10	14 30
siao ...	160 50	3 30
siarant ...	118 20	30 0
sibaccha ...	50 0	28 40
sibier r. op. ...	99 20	59 30
sibilia ...	22 0	35 50
sicby montes ...	115 0	58 30
sicilia ...	45 0	37 30
sidal ...	75 30	22 0
sidon ...	72 10	36 30
sierras de Penedall	47 20	30 40 *a.*
sierra neuada ...	298 0	49 0 *a.*
siete herman ...	170 0	9 30
sigesmel locus ...	41 0	11 30
sigistan r. ...	105 0	31 0
simiso ...	69 10	44 20
sina ...	70 0	41 40
sinai mons ...	75 0	30 0
sind ...	109 30	27 0
sindacui ...	165 10	55 40
sindam ...	99 0	32 20
sindiufu ...	142 10	44 0
singa ...	60 30	3 50 *a.*
singin ...	147 10	41 40
singui ...	155 30	55 30
singui ...	149 30	38 30
singuimatu ...	155 30	48 10
siminan ...	106 30	45 30
sinus Barbaricus...	74 0	4 0 *a.*
sinus S. Laurentij	325 0	49 0
sinus Mexicanus...	280 0	26 0
sinus Persicus ...	85 0	29 0
sione ...	59 10	12 40
sipanto ...	45 30	41 50

	Long.	Lat.
siquisita ...	312 40	19 50 *a.*
sirach ...	87 20	42 10
siras ...	90 40	30 40
sire ...	46 30	12 30
sirgiam ...	95 10	29 40
sissam Ins. ...	106 20	1 20
sistan ...	105 30	28 40
slaba ...	55 50	58 40
slauonia ...	47 0	45 0
slowoda ...	68 20	64 30
slowoda ...	86 30	58 50
slutzk ...	59 0	52 38
smacatlan ...	270 50	16 40
smirna ...	60 20	40 30
snauel ...	2 30	64 20
sobaha ...	63 10	16 10
socbasi fl. ...	108 0	48 50
soghgi ...	143 20	50 20
soha ...	92 30	23 50
solangi reg. ...	139 0	50 0
solidea ...	14 15	11 0
soloski ...	55 0	64 30
soltania ...	92 40	37 20
soram ...	86 50	35 40
sorand ...	351 40	61 0
sorlings ...	18 0	50 0
sosa ...	61 0	1 30
sossa fl. ...	108 30	64 30
sostan ...	117 40	28 0
spaam ...	96 50	33 30
spakado ...	46 50	45 20
spartivento ...	47 30	38 0
spier ...	35 30	49 20
spina ...	60 50	43 30
spicia ...	39 50	44 30
stachene ...	118 50	32 40
staci ...	94 0	30 40
stad ...	30 40	61 40
stadin fl. ...	306 20	50 0
staianfu ...	147 50	42 0
stagira ...	55 30	43 30
stampalio ...	59 50	36 40
stapholt ...	2 20	65 40
starabat ...	99 40	41 20
starigur ...	44 40	69 10
stecborg ...	42 30	58 50
stetin ...	42 10	53 50
stobi ...	52 30	44 0
stockolm ...	42 0	58 10
stoka ...	57 50	48 30
stolp ...	45 30	55 30
stora ...	35 50	35 40
stormesent ...	30 0	59 40
stornita ...	135 10	37 10
straight of Matu-chin	74 30	73 10
straun ...	84 30	43 30
strelna ...	79 40	61 10
streltze ...	79 40	62 0
strupuli cost ...	96 10	62 20
suachem ...	72 40	18 40
suastus fl. ...	119 20	35 0
sua zino ...	51 10	40 30
subao ...	153 50	10 0 *a.*
succuir ...	143 10	56 0
suedia reg. ...	40 0	60 0
suetinos ...	57 0	68 30
suffetuba ...	39 20	32 40
suguan ...	69 30	26 40
sumatra Ins. ...	134 0	0 0
sunda ...	138 0	6 40 *a.*
supa ...	156 30	1 10 *a.*
sus ...	87 30	34 30
susaca ...	73 40	48 0
susdal ...	74 20	56 40
swest ...	64 50	52 10
swinburne head ...	25 0	59 50
syr ...	90 30	21 30
syria ...	74 0	36 0
syracusæ ...	45 40	37 0
syrtis maior ...	48 30	29 30
syrtis minor ...	43 10	32 10
T.		
Tabaco ...	322 10	10 40
Tacan ...	152 20	48 50
Tacine ...	27 40	26 0
Tachnin fl. ...	125 50	62 0
Tacomiguo ...	72 40	4 0 *a.*
Tadelis ...	33 50	35 20
Tadinsu ...	156 20	49 40
Tagaranto ...	143 30	2 20
Tagaza ...	18 40	22 0
Taguima In. ...	154 30	5 20
Taiapura ...	142 30	1 50
Taigin ...	149 10	61 50
Taingu ...	152 0	63 30
Taiombara ...	144 20	4 0
Taiompura ...	145 50	1 30
Taiona ...	59 30	53 30
Talabora ...	312 0	26 20 *a.*
Talao ...	161 0	4 20
Talca ...	98 10	42 30
Talcan ...	85 0	47 0
Tamaco ...	270 15	24 40
Tamasa ...	75 30	46 0
Tambof ...	15 30	27 10
Tamos pr. ...	174 30	50 30
Tana ...	135 10	15 30
Tanamaibu ...	109 30	18 10
Tanchit ...	114 10	48 20
Tanes ...	30 50	35 20
Tapasipa ...	275 0	15 20
Tapuri mõtes ...	113 0	53 20
Tarama ...	301 15	13 15 *a.*
Taranto ...	48 0	40 30
Tarapaca ...	306 20	30 40 *a.*
Tarbacan ...	109 30	34 50

	Long.	Lat.
Targa reg.	32 0	25 0
Targa op.	31 20	23 40
Tarnassar	119 40	17 10
Tarragona	29 30	40 40
Tarso	71 20	40 0
Tartaho	162 40	38 40
Tartar	152 0	63 20
Tartaria reg.	130 0	62 0
Tasan	132 30	36 20
Tasica	66 40	22 10
Taskent reg.	129 0	49 0
Taskent op.	126 0	50 15
Taste	308 40	9 10
Tatracan	55 0	44 50
Tauasca	275 40	18 20
Tauais	108 40	42 20
Tauay	135 10	15 30
Tauest	49 20	63 30
Tauilla	18 10	37 20
Tauris	90 30	38 10
Taxila	121 40	34 50
Tebeld	41 10	10 10
Tebilbelt	23 10	29 30
Technaa fl	68 0	7 20 *a.*
Techort	35 50	27 10
Tefethne	16 10	30 0
Tega	47 20	25 30
Tegoram	29 30	30 0
Tegnat	27 40	28 10
Teient	17 0	30 30
Tellin	13 30	54 40
Temican	20 50	8 30
Teneriffe	8 10	27 30
Tendue op.	168 30	57 30
Tendue reg.	170 0	59 0
Tenesab	46 50	61 10
Tenlech	17 0	31 0
Teorregu	48 50	25 0
Tequandela	303 10	49 0
Tercera	358 20	39 0
Terenate	160 40	1 20
Terra alta	160 30	6 40 *a.*
Terra alta	45 20	15 20
Terra de los fumos	322 30	40 20 *a.*
Terra de Humos	348 40	1 30 *a.*
Primera Terra	172 10	0 30 *a.*
Terra de S. Vincente	346 40	2 0 *a.*
Tarsis	115 20	49 0
Tesebit	27 30	30 0
Thessalonica	53 40	44 20
Thesset reg. op.	20 0	29 10
Testigos	316 10	11 0
Teufar	37 30	27 10
Texir	11 30	22 30
Tezerin	24 50	30 40
Tezzeri	43 40	26 0
Thebet reg.	138 50	45 0
Thebet op.	138 50	44 0
Thene	79 20	37 40
Tholoman	144 20	40 0
Tholouse	28 40	43 50
Thomebamba	293 40	1 50 *a.*
Thunis	67 40	32 0
Thialso	49 40	22 40 *a.*
Tidore	160 40	0 40
Tigramahon	65 0	6 0
Tigris fl.	84 0	34 30
Timitri	133 10	49 20
Timocham	108 50	28 20
Tinazen	70 0	13 0
Tingui	155 0	43 0
Tinoca	166 0	32 0
Tinzu	164 0	48 40
Tipura	131 10	28 10
Tiguisana	305 20	16 0 *a.*
Tirna	47 0	49 0
Tisrich	95 20	28 10
Titicaca lacus	308 30	18 0 *a.*
Tochtepec	274 40	19 0
Tocros	54 50	46 0
Tosian	13 20	29 0
Togora	146 0	49 50
Takel	4 20	64 0
Toledo	22 20	39 40
Tollon	34 50	43 20
Tolometa	53 0	31 30
Tombute	20 50	15 0
Toram	134 30	7 0
Torn	47 0	53 10
Toropetz	62 40	57 50
Tortoza	29 30	40 30
Tortuga	303 50	20 20
Tortugas	312 20	10 40
Tosalis	143 40	37 30
Totoneac	248 20	36 0
Toul	33 10	49 10
Toure	27 30	47 50
Trabuco	56 30	31 30
Tranom Ins.	107 10	1 20
Tranooch	34 20	67 0
Trapam	43 30	37 30
Trapicari	305 10	7 0 *a.*
Trebizonda	74 30	44 40
Tremizen	29 0	34 10
Trent	40 10	46 10
Treta	68 0	37 20
Treuia	20 10	7 40
Triago Ins.	278 40	21 0
Tribanta	63 30	41 50
Tricalamata	120 10	7 30
Trier	34 10	49 50
Trieste	44 10	46 10
Trin	36 30	45 40
Trinidad	355 20	19 10 *a.*
Trinidad	295 50	21 20
Trinidad	319 20	9 0
Trinitie harbor	308 30	36 0

	Long.	Lat.
Tripolis antiqua ...	44 20	30 20
Tripoli de Barbaria	45 20	30 30
Tripolis Soriæ ...	72 20	37 0
Troia ...	59 0	42 30
Troy ...	31 0	48 10
Tuat ...	29 0	28 30
Tuban ...	154 10	5 40 *a.*
Tucare ...	32 20	7 0
Tucca ...	38 10	33 20
Tucken ...	51 30	57 40
Tuesa ...	81 15	18 0
Tugasar ...	16 40	14 30
Tuia ...	82 50	52 0
Tulla ...	72 0	53 20
Tumbes ...	291 40	4 10 *a.*
Tumboblanco ...	294 0	3 0 *a.*
Tumena ...	90 50	29 30
Tamisa ...	84 10	24 0
Tuna ...	41 50	64 30
Tunei ...	72 10	9 40
Tunis r. op. ...	40 0	36 0
Turbet ...	99 50	34 0
Turchestan reg. ...	110 0	47 0
Turfon ...	131 30	56 30
Turris lapidea mons	125 0	47 0
Turses ...	103 40	34 0
Tursis ...	103 40	37 30
Turunbaia ...	76 20	24 50 *a.*
Tutega ...	17 0	6 30
Twer ...	68 10	57 10
Tybi ...	91 50	19 40
Tyrus ...	71 35	35 30
Tzercas ...	79 50	49 20
V.		
Vadi ...	54 40	16 0 *a.*
Vahuliez ...	90 40	60 50
Vaigirmale ...	119 0	18 0
Vaigui ...	150 50	39 0
Val Parayso ...	300 0	33 0 *a.*
Valderas ...	261 50	22 30
Valentia ...	29 20	39 40
Valunta ...	56 0	27 50 *a.*
Vamba fl. ...	49 40	5 0 *a.*
Van ...	86 30	36 50
Vangue ...	48 40	8 50
Var ...	120 30	22 40
Varcano ...	107 50	39 0
Varon ...	83 30	70 30
Varta ...	46 50	51 40
Vasianar ...	75 40	49 0
Vastan ...	85 10	36 50
Vasten ...	39 30	59 50
Vatacaba ...	53 40	12 30 *a.*
Vaygath Ins. ...	81 30	69 20
Vban ...	96 50	32 0
Vche ...	110 40	31 30
Vekelax ...	54 20	62 0

	Long.	Lat.
Venetiæ ...	41 40	45 50
Vella ...	77 0	13 0
Verdiso ...	59 50	45 0
Verdum ...	32 10	49 20
Verma r. ...	133 0	21 30
Verma op. ...	130 20	20 10
Verona ...	40 40	45 50
Vertoplate ...	130 30	1 30 *a.*
Vesgirt ...	116 20	41 30
Vguin ...	161 10	39 20
Viana ...	17 30	42 0
Viatca ...	87 50	59 30
Vich ...	81 40	53 50
Videpski ...	59 0	57 0
Vienna ...	45 30	48 30
Vigangara ...	80 40	14 40 *a.*
Villac ...	48 0	46 50
Villa longa ...	28 20	7 40
Villa Conde ...	17 30	41 30
Vilna ...	54 30	55 0
Vindius Mons ...	124 0	28 0
Virgines ...	178 40	1 20
Virginia ...	302 0	36 0
Visigrod ...	61 30	51 30
Bona Vista ...	4 30	15 30
Buena Vista ...	308 40	40 10
Buena Vista ...	177 30	13 30
Viterbo ...	41 50	42 40
Vkkil ...	53 10	57 0
Vllao ...	242 10	30 30
Vllao ...	240 30	21 0
Vlm ...	37 50	48 50
Vocam ...	116 8	39 0
Vociam ...	128 0	40 0
Volga fl. ...	75 40	58 0
Vpsalia ...	42 50	60 0
Vque ...	60 40	6 40 *a.*
Vraba ...	297 20	7 30
Vraba ...	285 30	10 40
Vrcamia ...	23 50	46 0
Vrcos ...	301 0	14 50 *a.*
Vrdubar ...	90 30	37 0
Vrgis fl. ...	85 50	53 20
Vristigna ...	38 40	39 40
Vsargala Mons ...	32 50	27 0
Vstiga ...	43 15	39 0
Vstiug ...	79 30	61 30
Vstuzna ...	67 0	59 20
Vtual ...	42 40	62 50
W.		
Waesbergen ...	39 0	57 30
Wardhuys ...	50 30	70 30
E. Warwickes fore-land	323 10	62 0
Coun. Warwicks sound	330 40	64 40
Wassilgorod ...	81 50	56 40
Woxen ...	49 20	52 30

	Long.	Lat.
Weimought	23 50	51 0
Welichi	96 30	56 0
Weliki poyassa	101 20	63 30
Weliki tumen	95 40	56 20
Welisz	63 40	56 50
Weroy	36 50	68 40
Wesel	31 30	51 30
Westerhol	40 30	67 40
Whitbe	24 30	55 0
Wiborgh	56 30	62 35
Wiesma	67 30	55 0
Wight	25 10	50 30
Sir Hugh Willowbies land	55 0	75 0
Winterton	27 30	53 30
Winerus	18 40	43 40
Wococan	307 30	34 0
Wologda	73 50	59 30
Wologda	74 30	60 0
Wolsk	68 30	55 50
X.		
Xaiel	85 30	15 40
Xandu	168 40	55 40
Xanes	311 30	11 0
Xagnes	282 0	20 30
Xara	130 0	17 0
Xibuar	116 0	46 30
Xinxa	301 30	12 0 *a.*
Xumete	304 20	23 0
Y.		
Yermouth	27 30	53 0
Yorck	23 30	54 30
Yuagua	303 30	21 0
Yuchcope	22 50	56 30
Z.		
Zabe	67 20	5 30 *a.*
Zacabedera	140 40	13 10
Zacana fl.	60 40	13 0 *a.*
Zacatula	269 40	20 0
Zachabirtenduc	165 10	58 30
Zachet	76 40	6 0
Zacatora Ins. & op.	88 0	12 50
Zagatray	105 0	45 0
Zahaspa	101 20	42 30
Zahu	141 20	28 0
Zaiton	157 30	28 0
Zalines	51 50	58 30
Zama	49 30	14 0 *a.*
Zama	74 40	11 40
Zambere fl.	55 0	19 10 *a.*
Zamfara	41 0	16 0
Zamilla	89 0	28 20
Zanhaga reg.	20 0	24 0
Zanzibar	73 50	6 30 *a.*
Zaphalonia	52 0	38 30
Zara	46 25	45 40
Zaradrus fl.	125 0	94 0
Zardadain	143 10	32 20
Zarim	135 40	14 40
Zauon	41 30	50 0
Zazela	81 40	7 40
Zerbeng	138 40	35 40
Zebil mons	47 0	17 0 *a.*
Zedica	48 0	29 30
Zegzeg r. op.	36 40	14 40
Zeila	80 0	11 0
Zeit	77 0	5 0
Zembere lac	55 0	11 30 *a.*
Noua zemla	83 30	74 0
Zengian	158 20	37 20
Zerigo	56 0	36 0
Zerzer	79 0	17 50
Zet	53 0	17 10 *a.*
Zibit	70 0	22 10
Zigeck	45 50	40 50
Zigide	55 0	10 40
Zil	115 0	15 0
Zimbaos	59 0	25 20 *a.*
Zimbro	50 50	22 40
Zingis	76 10	49 30
Zire	107 10	30 10
Ziz	27 0	26 30
Zodaha	143 30	8 20 *a.*
Zodiala	57 50	4
Zordalanel	137 30	
Zophal	64 20	
Zoquila	58 30	
Zuenziga r.	25 0	23
Zuiatzko	85 20	56 0
Zunbal	39 30	37 30

FINIS.

BIOGRAPHICAL INDEX

OF

NAMES IN THE "TRACTATUS DE GLOBIS" OF ROBERT HUES.

PAGE

Abulfeda—On the measurement of a degree by order of Almamun ... 92

Length of a parasang 93

Place whence the Arabians reckoned their longitude 96

This Arabian historian and geographer belonged to the same family as the famous Saladin, and was one of the Ayubites who reigned at Hamal in Syria. He was born in 1273, and died in 1331. He took part in the wars which resulted in the complete extirpation of the colonies formed by the Crusaders in the East, and in the wars of the Sultans of Egypt and Syria against the Mongols. His works are the *Universal Chronicle* and the *Geography*. The latter work contains an account of the system of the sphere as then understood in the East, tables of latitude and longitude, and detailed descriptions of countries and seas. The first complete edition of the works of Abulfeda was published by Renaud (Paris) in 1840, with a French translation.

Achæus—Number of the Hyades according to 56

Achæus, of Eretria in Eubœa, was born B.C. 484, the contemporary of Sophocles and Euripides. The titles of ten of his tragedies and of seven of his satirical dramas are known, but only fragments have been preserved, collected and edited by Urlichs (Bonn, 1834).

Agenor—Father of Europa 77

He was son of Poseidon and Libya, and King of Phœnicia, twin-brother of Belus.

Albarenus—Position of α *Arietis* in the time of 29

Albategnius—On the length of the year 27

His real name was Muhammad ibn Jafar ibn Senan Abu Abdallah, known as Albatenius and Albategnius. He was born in the ninth century, at *Baten*, near Haran in Mesapotamia, whence his name of Al-*baten*-ius, and died in 929 A.D. His observations were taken between 877 and 918, at Rekbah on the Euphrates, and at Antioch. His chief work was translated into Latin under the title of *De Scientia Stellarum*, and printed at Nuremberg, 1537, and at Bologna, 1545, with a commentary by Regiomontanus. It showed that the Arabs had tables which gave the altitude of the

Ptolemy and Euclid, and a treatise on optics, and on twilight; translated into Latin, and published at Basle in 1572. It was in accordance with his ideas that the first spectacles were made.

Abul-Abbas-Abdallah-al-mamun, the Abbasside Khalîfa, was born at Baghdad in 786, and died in 834. He was son of the celebrated Khalîfa Harun-al-rashid, in whose life-time he administered the Persian province of Khorasan. He succeeded in 813. His reign was a period of progress and civilization. He caused numerous Greek scientific and philosophic works to be translated into Arabic, and especially fostered the study of mathematics and astronomy. He founded an observatory at Baghdad, and caused a degree of the meridian to be measured on the plain of Mesopotamia. His chief astronomers were Albategnius, Albumazar, the Jew Maschallah, and the Persian Abdallah-ibn-Sehl.

Alfonso X (el Sabio), King of Castille and Leon, was born in 1226, and died in 1284. He was brother-in-law of Edward I of England, and succeeded his father, San Fernando III, in 1252. This king cultivated the science of the Arabs of Spain, and was devoted to literary pursuits. An unwise and vacillating politician, he was an able lawgiver and a great patron of literature. He founded the University of Salamanca, promoted the study of the Spanish language, and compiled a code of laws. The astronomical tables prepared under his auspices were in universal use until the beginning of the sixteenth century. They were called the Alfonsine Tables, and were probably the work of Arabian astronomers of Granada, who lived at the court of Alfonso. The tables are dated 30 May 1252, and were first printed at Venice in 1492. The room is still shown in the alcazar of Segovia, where Alfonso studied astronomy. His code of laws was called "Las Siete Partidas", and was almost entirely the king's own work. The celebrated *Cronica de España*, a history of Spain from the earliest times to the death of his father, is also attributed to Alfonso X.

latter part of his life at the court of Antigonus Gonatus, King of Macedonia. The first poem, called *Phenomena*, consists of 732 verses, the second, *Prognostica*, of 422. These poems are believed to be versified editions of two works by Eudoxus, which are lost. The positions of the constellations and the path of the sun in the zodiac are described. The opening verses contain the passage quoted by St. Paul (*Acts* xvii, 28), " For in him we live, and move, and have our being, as certain also of your own poets have said." The poems were very popular, and there were several Latin translations.

Archimedes—Improved the globle or sphere 5
Discovered the proportion between diameter and circumference of a circle 12
Every liquid body at rest has a spherical surface 14
Adopted the length of the year from Calippus 26
Cites Aristarchus as to the sun's apparent diameter 89

Born 287 B.C. A kinsman, certainly a friend of Hiero, King of Syracuse. The most famous mathematician of ancient times. He studied, in Alexandria, under Conon, and then returned to Syracuse. He constructed various engines of war for Hiero, which were used when Marcellus besieged the town, and long delayed its capture. He is said to have set the Roman ships on fire with a burning-glass. He built a large ship, and moved it into the sea by means of a screw, being a present from Hiero to Ptolemy, King of Egypt. He also invented a water-screw for pumping water out of the ship's hold. He constructed a kind of orrery for representing the movements of the heavenly bodies. He discovered the proportion between the circumference and diameter of a circle ; and many other solutions of mathematical problems. He was killed by Roman soldiers when Marcellus took Syracuse.

Arias Montanus—Translation of **Benjamin of Tudela**. Error respecting Canopus 61
Name of Abyssines 78

Arias Montanus, a learned Spaniard, was born in 1527, and died in 1598. He was a great linguist, and travelled over every part of Europe. He also accompanied the Bishop of Segovia to the Council of Trent. He had charge of the publication of a new edition of the Polyglot Bible (1572), and Philip II offered him a bishopric, which he declined. His translation of Benjamin of Tudela is in Latin.

Aristarchus, Samius—Followed Calippus, in calculating the length of the year 26
Calculation of the sun's apparent diameter 89

He flourished at Samos in 270 B.C. It occurred to him that the illumination of the moon by the sun afforded a means of estimating the sun's distance. He estimated the sun's distance at nineteen times that of the moon, or a twentieth of its true value.

By ascertaining the exact time between new moon and half full moon he got two angles in a triangle, one side of which is the distance required. None of his works remain, except the treatise on the distances of the sun and moon.

Aristotle—On the height of Mount Athos 8
Height of Caucasus celebrated by 9
Reported that Sesostris gave up his scheme of uniting the Mediterranean and Arabian Seas, because the surface of the latter was higher than the former 14
Terminated the zones with the tropics and Arctic and Antarctic circles 38
His calculation of the circumference of the earth 80

He was born at Stageira, a seaport in the district of Chalcis. Born 384 B.C. His father, who was a physician, introduced him to the court of the King of Macedonia. On his father's death he went to Athens, and became a disciple of Plato. He lived at Athens for twenty years. He then accepted an invitation of Philip of Macedon to become the tutor to his son Alexander. Stageira, which had been destroyed by Philip, was rebuilt at the request of Aristotle, and a grove was planted there, for himself and his pupils. Here he lived with his royal pupil for four years. In 335 Aristotle returned to Athens, and delivered his lectures to his disciples, while walking in the groves which surrounded the lyceum. He died at Chalcis in 322 B.C., aged 63. His works were studied by the Arabian men of learning, led by Avicenna and Averrhoes, and, through the commentaries of St. Thomas Aquinas, at the universities of Paris and Oxford. In the fifteenth and sixteenth centuries the editions of Aristotle were very numerous.

Artemidorus—On the apparent size of the sun at his setting ... 64

A Greek geographer of Ephesus who flourished about B.C. 100. He was also a great traveller, but his work, which was valued highly by the ancients, is lost. An abridgment was made by Marcianus of Heracleia, and fragments of this abridgment have been preserved. Artemidorus is frequently quoted by Strabo and Pliny.

Arzachel—On the length of the year 27
On the time of the vernal equinox 28
Position of *a Arietis* 29
Distance of the tropics from the equator 32

A celebrated Jewish astronomer of Toledo, living about 1080 A.D. He determined the apogee of the sun by 400 observations, and fixed the obliquity of the ecliptic at 23° 34′. Arzachel is the author of the Tables of Toledo, which probably served as a basis of the Alphonsine Tables.

Atlas of Libyra—Said to have invented the globe or sphere ... 5

Son of Japetus and Clymene, according to Hesiod; who said that he bore up heaven with his head and hands. He is described

A grammarian and poet, born at Cyrene, chief librarian at Alexandria under Ptolemy Philadelphus, B.C. 260 to 240, when he died. The titles of forty of his works are known to us, but the fragments that have been preserved are chiefly poetical. They consist of six hymns, seventy-three epigrams, and parts of elegies. Catullus imitated one, in his *De Coma Berenices.* His prose works are entirely lost.

Francisco Campano, born in Tuscany, and Secretary to Cosmo de Medici. He was a classical scholar of eminence.

Geronimo Cardan, a celebrated Italian physician and philosopher, was born at Pavia in 1501, and died at Rome in 1576. He was educated at Venice and Padua, and settled at Milan as a physician. In 1552 he visited Scotland at the invitation of John Hamilton, Archbishop of St. Andrew's, and saw King Edward VI in London, on his way back to Italy. He was unhappy in his family relations, his wife being a scold; one son was beheaded for poisoning his wife, and the other was so incorrigible that Cardan was obliged to disinherit him. His extraordinary life is related by himself in his *Vita propria.* His best-known work is entitled *De Subtilitate,* which was vigorously attacked by Scaliger. This and the *De Rerum Varietate* comprises all the knowledge Cardan had acquired in medicine and natural history, most of his ideas being borrowed from Aristotle and Pliny. But he wrote upwards of 222 other treatises.

Consul B.C. 60. He died in B.C. 59, the year of Cæsar's Consulship.

Censorinus wrote a book called *De Die Natali,* in 238 A.D. It treats of the generation of man, his natal hour, the influence of the stars on his career, and the various methods for the division and calculation of time. He was a native of Rome, but nothing is known of him.

Jacob Christman, a learned German, was born at Johannisberg in 1554, and died in 1613. He knew Arabic, Syriac, Hebrew, Chaldee, Greek, Latin, French, Italian, and Spanish. He travelled for some years, and eventually settled at Heidelberg, where he taught the Eastern languages and logic for thirty years. His work on the astronomy of Alfraganus, with a commentary on the calendars (Frankfort, 1590), is the one referred to by Hues.

The date of the work of Cleomedes on the *Circular Theory of the Heavenly Bodies* is uncertain, and nothing is known of the writer. It treats of the universe, of the zones, motions of the stars and planets, and of the magnitude and figure of the earth. He gives the only extant account of the way in which Eratosthenes and Posidonius attempted to measure an arc of the meridian. It is probable that Cleomedes flourished before Ptolemy. Sir Edward Bunbury looks upon the work of Cleomedes as an epitome of the views of Posidonius.

Daughter of Ptolemy Auletes, born B.C. 69. Dethroned by her brother; she was restored by Julius Cæsar, who loaded her with honours, and induced her to come to Rome. On Cæsar's death she fled to Egypt. Meeting Mark Anthony in Cilicia she entirely captivated him, and they returned to Egypt together. At the battle of Actium she fled with her fleet, and was joined by Anthony at Alexandria. It was then that she formed the plan of transporting her fleet into the Red Sea. She betrayed Anthony, who, however, died in her arms, and finding no favour from Augustus, she poisoned herself by the bite of an asp—B.C. 30, aged 39—the last of the Ptolomies.

Cleostratus Tenedius—First divided the zodiac into signs, according to Pliny 24

First observed the configuration of the *Hædi* and *Capella* ... 54

An astronomer of Tenedos, inventor of the cycle of eight years (used before Meton introduced the nineteen year cycle), according to Censorinus. Lived B.C. 548 to 432. It is Higinus who says that Cleostratus first observed the Hædi.

Coignet, Michael—His error respecting the height of the poles in rumb sailing 131

Exposed the mistake of those who thought that the rumbs met at the poles 131

Columella—On the cause of the solstices 25

L. Junius Moderatus Columella was the most important of all the Roman writers on rural affairs. He flourished in the first half of the first century after Christ, and was a native of Cadiz, but generally lived at Rome. His work is a comprehensive treatise on agriculture, in twelve books.

Conon—The Alexandrian mathematician. Constellation of Berenice's Hair 57

A native of Samos, friend and pupil of Archimedes. His works are all lost, but his observations are referred to by Ptolemy. Seneca tells us that he made a collection of the solar eclipses observed by the Egyptians. His naming of the constellation of Berenice's Hair is on the authority of the poem of Callimachus, translated by Catullus.

Copernicus—On the length of the year 27

Position of *a Arietis* 29

Distance of the tropics from the equator 32

Censured by Scaliger 47

His enumeration of the stars 49

Reckoned the longitude of stars from *a Arietis* 50

Erroneous estimate of the width of the isthmus between the Mediterranean and Arabian Sea 71

Apparent diameter of the sun 90

Nicolaus Copernick (Copernicus) was born at Thorn in Poland, in January 1472. He was educated at the University of Cracow, and was afterwards some time at Bologna, where he studied mathematics and astronomy. In about 1500 he settled at Rome, and eventually returned to his native country. He became a canon of Frauenberg near Danzig, in the diocese of Wermland, where his uncle was bishop. His criticisms of the Ptolemaic system, and the work describing his own theory, were completed in 1530, but not published till 1543 at Nuremberg, being dedicated to Pope Paul III. He placed the sun in the centre of the universe, and he correctly explained the variation of the seasons and the precession of the equinoxes. Copernicus died at Frauenberg in the year that his work was published. It was entitled *De revolutionibus Orbium Cœlestium.*

Cneius Domitius Corbulo, brother-in-law of the Emperor Caligula, was Consul A.D. 39. He commanded an army in Germany with great success in the reign of Claudius, and Nero entrusted him with the supreme command against the Parthians. He conquered Armenia, and was always faithful to Nero, who condemned him to death A.D. 67. On receiving the news, he committed suicide.

Contemporary of Cicero and Catullus. He was probably a native of Verona, and died during the reign of Augustus. All his works are lost, but in 1471 a volume was published at Venice containing a series of biographies of nineteen Greek and Roman generals, attributed to one Probus. Probably the biographies were written by Cornelius Nepos, and abbreviated some centuries afterwards by Probus.

Andrea Corsali, an Italian navigator, was born at Florence, and entered the service of King Emanuel of Portugal. He received command of a ship in which he made a voyage to India and China, and visited Muscat and part of Persia. The narrative of his voyage is contained in two letters to Lorenzo de Medici, dated 1515 and 1517. Ramusio inserted it in his collection of voyages.

Crates of Mallus, in Cilicia, was famous as a grammarian who lived at Pergamus under the patronage of Eumenes and Attalus II B.C. 160. He also wrote a commentary on the Theogony of Hesiod, and a work on geography, of which only a few fragments remain.

A Greek historian who was sent as ambassador to India in about 312 B.C. Strabo names him as one who spread false and fabulous reports about India. His work is lost.

Demetrius I of Macedonia (Poliorcetes), son of Antigonus. He succeeded his father in B.C. 301, when the latter was slain in the battle of Ipsus. In B.C. 286 he fell into the hands of Seleucus,

King of Syria, and died in captivity. He was a man remarkable for activity of mind, fertility of resource, and promptitude in the execution of his schemes.

Democritus—On the length of the year 26

Dicæarchus—On the height of mountains 11

Dicæarchus, a philosopher contemporary with Aristotle, was born at Messina, but passed his life in Greece. He died about B.C. 285. His works were partly geographical and partly historical, but they are all lost except a few fragments. One of his works was *On the Height of Mountains*, mentioned by Pliny.

Diodorus Siculus—Statement that Atlas of Libya discovered the use of the globe 5

A contemporary of Cæsar and Augustus; born at Agyrium in Sicily. He made it the business of his life to write a universal history, and for this purpose he travelled much, and was for several years at Rome, collecting materials. He wrote in about B.C. 8. The work consisted of forty books, of which only fourteen have been preserved.

Dion—His error respecting the length of the year 27

Dion Cassius was born at Nice in Bithynia in 155 A.D. He was carefully educated, and came to Rome soon after the death of the Emperor Marcus Aurelius. He became a Senator and voted for Pertinax on the death of Commodus. During the reign of Severus he retired to Capua to write his history. Consul A.D. 220. Proconsul in Africa and Pannonia. Under Alexander Severus he was again Consul A.D. 229. He retired to Nice, where he completed his history and died. An important portion of his work has been preserved.

Dionysodorus—An epistle to the gods, on the earth's semidiameter, found in his tomb 81

A Greek geometer of Cydnus. The date of his life is uncertain; but according to Pliny a letter was found in his tomb, addressed to the living. In it he declared that the radius of the earth was 42,000 stadia. This is the most exact measurement recorded by the ancients. 42,000 stadia is equal to 7,770 kilomètres.

Dionysius Afer—On the height of Atlas 11

Shape of the earth compared to a hand 68

Placed Taprobane under the tropic of Cancer 76

Dionysius Exiguus—A Roman Abbot. Introduced the use of letters to designate days in the Calendar 21

Dionysius Periegetes—Absurd height given to the Pillars of Hercules 9

Author of *Periegesis*, describing the earth in hexameter verse. This work is still extant, and was very popular in ancient times. He probably flourished in the beginning of the fourth century; from the reign of Nero to Trajan. The work merely professes to be a summary.

Drake, Sir Francis—His voyage of circumnavigation ... 3, 15

Endymion—First observed the phases of the moon, according to Pliny 46

A youth beloved by Selene, who was granted the boon of eternal sleep on Mount Latmus; kissed by the soft rays of the moon.

Eratosthenes—Harshly censured by Strabo 1

On the earth being shaped like a globe, with some irregularities ... 6

Height of the atmosphere 10

Height of mountains 12

Irregularities in the surface of the sea 13

Length of a degree in furlongs 31

A narrow zone on the equator held to be habitable 38

Showed that the number of uninhabitable zones was erroneous ... 39

Commentaries on Aratus attributed to 49

View as to the shape of the earth 68

Believed that Europe was once joined to Africa 70

The isthmus between the Mediterranean and Arabian Sea, once submerged 71

On the circumference of the earth 80

Distance between Syene and Alexandria 81

Received the report of distances without actual measurement ... 84

Errors in his distances 84

Observation of the distance from Rhodes to Alexandria ... 88

Erroneous calculation respecting the earth's circumference 90, 94

Eratosthenes of Cyrene was born B.C. 276. Leaving Athens at the invitation of Ptolemy Euergetes, he was placed over the library at Alexandria. He died B.C. 196, aged eighty, of voluntary starvation, having lost his sight, and being tired of life. He made the distance of the tropics from the equator to be 23° 51′20″; which was adopted by Hipparchus and Ptolemy. His great work was an attempt to measure the magnitude of the earth. He assumed that Syene (Assouan) was on the tropic, because he was told that vertical objects cast no shadow there, on the day of the summer solstice. He also assumed that it was in the same longitude as Alexandria, in which he was 3° out. In determining the latitude of Alexandria he used the hemispherical dial of Berosus, and so obtained the arc between Alexandria and Syene. The result was 250,000 stadia for the circumference of the earth. He systematised the scattered geographical information then existing, and combined it in a great work, which is unfortunately lost. We only have fragments quoted by later writers.

Etesias—Indian histories. Apparent size of the sun in India ... 64

Euctemon—Time of the solstice in his time 28

An astronomer who worked with Meton. Ptolemy refers to him as an authority on the rising and setting of stars.

Eudoxus—Advance of knowledge since his time 1

Stars called by the same names in his time 49

Wonderful account of the falls between the Caspian and Scythian Seas 69

This learned Frisian was born at Dokkum in 1508, and died at Louvain in 1555. In 1541 he became Professor of Medicine at Louvain, but his principal works are on mathematical and astronomical subjects. His *Methodus Arithmaticæ Practicæ* appeared at Antwerp in 1540 ; *Totius Orbis Descriptio* (Louvain, 1540) ; *De Principiis Astronomiæ* (Paris, 1547) ; *De Usu Annuli Astronomica* (Antwerp, 1558) ; *De Astrolabio Catholico et usu ejusdem* (Antwerp, 1556). He also edited the *Cosmographia* of Appianus.

His remains of a Latin translation of Aratus are in verse, and critics have denied his authorship without sufficient reason. The *scholia* appended to the translation are attributed to Cæsius Bassus. The military exploits of Germanicus are recorded by Tacitus. His mother, Antonia, was a niece of the Emperor Augustus ; his father, Nero Claudius Drusus, was son of the Empress Livia, and brother of Tiberius ; so that he was brother of the Emperor Claudius. He was born B.C. 15, and died A.D. 20.

Hugo Grotius was born at Delft in 1583, and died at Rostock in 1645. He was a statesman and a writer on many subjects. The work referred to by Hues is his *Syntagma Aratæorum Græce et Latine cum notis* (Leyden, 1600).

He reigned from A.D. 117 to 138, and was born A.D. 76 at Rome.

His *Periplus* has been preserved, being a Greek translation of the Punic original. The date of the voyage is about 500 B.C. The object of the voyage of Hanno was not discovery but colonisation. The expedition consisted of sixty ships, and a great number of men and women. The first settlement was formed on the coast two days' sail beyond the Pillars of Hercules. Some days afterwards they founded five other towns on the African coast. They passed the mouths of the great rivers, and came to a country where

Naplius—Constellation of the Great Bear, said by Theon to have been invented by 50

Nearchus—His history contemned by Strabo 1

A native of Crete. Alexander gave him the chief command of the fleet which descended the Indus, and navigated the Persian Gulf, B.C. 325. After Alexander's death he joined the fortunes of Antigonus. His work is lost, but its substance has been preserved by Arrian.

Nicander—Gives the whole bull in delineating the constellation ... 48

Places the *Pleiades* in the tail of the bull 56

A Greek physician and grammarian, as well as a poet, who flourished B.C. 185 to 135. His poems mainly treat of venomous animals and the wounds inflicted by them, of poisons and antidotes, and are full of absurd fables.

Nicias (in **Eustethius**), derived the name of Europe from Europus ... 77

Nonius, Petrus—Proved the heights of mountains reported by the ancients, to be fabulous 9

Pointed out the error, in interpreting Hipparchus, as to the position of *Polaris* 30

On the length of the furlong 83

On the circumference of the earth 88

Held the Arabian mile to be the same as the Italian 93

On the length of twilight 113

Believed that the magnet was weakened by long use 120

On rumbes and the practice of using them on the globe ... 127

Held that the rumbes consisted of portions of great circles ... 130

The rumbes do not enter the poles 131

The great Portuguese mathematician and astronomer. Pedro Nunez (or Nonius) was born at Alcazar in Portugal, in 1497. His work, *De Arte et Ratione Navigandi*, was published at Basle in 1567. He was the first cosmographer who exposed the errors of the plane chart, and he gave the solutions of several astronomical problems, including the determination of the latitude by sun's double altitude. His treatise on algebra was printed at Antwerp in 1567. Nunez, who was Professor of Mathematics at the University of Coimbra, died in 1577, aged 80.

Numenius—General to Antiochus. His victory by sea and land, on the same spot 72

Oenopides—Length of the year 26

An astronomer of Chios, who derived most of his knowledge from the priests of Egypt. He is said to have invented a cycle of fifty-nine years for bringing the lunar and solar years into accordance. The date of his life is uncertain.

Orontius Finæus—Difference of longitude by the motion of the moon 97

Ortelius, Abraham—Excellence of his geographical tables ... 17

Oxford, Monk of—Explored the polar regions by his skill in magic ... 10

Parmenides—His extension of the torrid zone 38

Pliny was born A.D. 23, and died, aged 56, A.D. 79. He was born either at Verona or at Como. He went to Rome when quite young to receive his education, and served in the army in Germany. During the reign of Nero he lived in retirement, and in A.D. 71 he went to Spain as Procurator, becoming guardian to his nephew, the younger Pliny, at about the same time. In 73 he returned to Rome, during the reign of Vespasian, whom he had known in Germany, and he now became one of that Emperor's most intimate friends, as well as the friend of his son Titus. He devoted nearly his whole time, during many years, to study, and amassed a vast amount of information, leaving to his nephew 160 volumes of notes. A.D. 77 he completed his *Historia Naturalis*, dedicated to Titus. He was appointed Admiral by Vespasian, and in 79 A.D. was at Misenum with the fleet when the eruption of Vesuvius took place. Approaching too near to observe the phenomena he was suffocated. Pliny was a mere compiler, without originality, or even the ability of sifting and arranging his materials.

The *Historia Naturalis* is divided into thirty-six books, besides the dedication to Titus, table of contents and list of authorities. The next book is the one in which he treats of the heavenly bodies, and of the physical conditions of the earth, and his historical notices of the progress of astronomy are very valuable. The four following books are devoted to geography.

His nephew, Pliny the younger, filled numerous important offices, was an orator, a learned scholar, and the intimate friend of Tacitus. His extant works consist of a eulogy of Trajan, and ten books of letters, which furnish materials for his life and notices of his contemporaries. He was born A.D. 62, but the time of his death is unknown. He gave an account of the circumstances of his uncle's death in a letter to Tacitus, but the most valuable and interesting letters are included in his correspondence with the Emperor Trajan.

The biographer was born at Chæroneia in Bœotia, and was a young man when Nero visited Greece. He lectured at Rome, and is said to have been the preceptor of Trajan, but passed the latter part of his life in his native town. The time of his death is unknown. His parallel lives of forty-six Greeks and Romans, arranged in pairs, have immortalized his name. His lives of the five first Roman Emperors and of Vitellius are lost.

This historian was a native of Megalopolis in Arcadia, born about 204 B.C. His father, Lycortas, was one of the most distinguished men of the Achæan league, who trained him in political knowledge and the military art. When Greece was conquered by the Romans, Polybius was taken to Rome, where he became the friend of Scipio, and he was present at the fall of Carthage. Returning to Greece, he interceded successfully with the Romans for lenient treatment of his countrymen. He travelled extensively, collecting materials for his history, and died B.C. 122. The greater part of his work is lost, only the first five books being complete.

A Greek historian, native of Larissa, and author of a history of Alexander the Great. His work is lost, but he is often quoted by Strabo.

Cn. Pompeius Magnus was born B.C. 106. He assumed command of the war against Mithridates B.C. 66, and completed his Eastern conquests in B.C. 63 ; B.C. 48 he was defeated by Cæsar in the battle of Pharsalia, and was assassinated when he was in the act of landing on the coast of Egypt.

A Stoic philosopher, native of Apameia in Syria, born about B.C. 135. He went to Athens, travelled extensively, and fixed his abode at Rhodes. He went as Ambassador to Rome, B.C. 86, and

Claudius Ptolemy observed at Alexandria in A.D. 139, and was alive in A.D. 161. Nothing more is known of him personally. His great work (Μεγαλη Συνταξις) or Μεγιστη, was known to the Arab translators as the *Almagest*. It was first printed at Venice in 1496 from a full epitome begun by Purbach, and finished by Regiomontanus. The first complete edition appeared at Venice in 1515, made from the Arabic. The version of George of Trebizond, made from the Greek, was published in 1528. The first Greek text was published at Basle in 1538. The catalogue of stars was published at Cologne in 1537, with forty-eight drawings of the constellations by Albert Dürer.

The first book of the *Almagest* treats of the relations of the earth and heaven, the theory of the sun and moon, and the sphere of the fixed stars and planets. It also contains an account of the observations proving the obliquity of the ecliptic, and geometry and trigonometry enough for the determination of the connection between the sun's right ascension, declination, and longitude, and for the formation of a table of declinations to each degree of longitude. The second book is on determination of latitude, the points at which the sun is vertical, the equinoctial and solstitial shadows of the gnomon, with several tables. The third book is on the length of the year, and on the theory of solar motion. The fourth and fifth books are on the theory of the moon, and the sixth on eclipses. The seventh and eighth books are devoted to the stars. The catalogue gives the longitudes and latitudes of 1,022 stars, described by their positions in the constellations. The remainder of the thirteen books is devoted to the planets.

Ptolemy was largely indebted to Hipparchus for his materials, and for his methods of calculating and observing.

Ptolemy's geographical syntaxis is a catalogue of names of places, with their estimated latitudes and longitudes, forming the materials for his map of the known world. It maintained its position as the accepted geographical text-book until the fifteenth century without a rival. The treatise of Ptolemy was based on the earlier work of Marinus of Tyre. Ptolemy assumed the earth to be a sphere, but the mode of laying down positions by imagining great circles passing through the poles called meridians, and other circles, one of which was a great circle equidistant from the poles, and the others parallel smaller circles, had been established from the time of Eratosthenes, as well as the division of great circles into 360°. But Ptolemy introduced the terms longitude (μηκος) and latitude (πλατος), and the plan of designating the positions of places by stating the numbers which represent the latitudes and longitudes of each. He divided his degrees into twelfths. His division of the earth into zones, which he called climates, was made with reference to the length of the longest day in each.

The *Geographia* of Ptolemy was printed at Rome in 1462, 1475, 1478, 1482, 1486, 1490; the editions of 1482 and 1490 being the best.

George Purbach was born at Linz in 1423. He was Professor of Astronomy at Vienna, where he constructed astronomical instruments. He commenced a translation of Ptolemy, and wrote on the theory of the planets, and on eclipses. Purbach died at Vienna in 1461.

The famous philosopher was probably born at Samos in about B.C. 608, or according to others in B.C. 570. He is believed to have travelled extensively; to have visited Babylon, and to have studied in Egypt. He eventually settled at Crotona, a Greek colony in the south of Italy, and there established a club or society for the study of the master's religious and philosophical theories. Pythagoras taught the doctrine of transmigration of souls, and made considerable advances in mathematical science, but his teachings were kept secret by the brotherhood into which his disciples were formed. Eventually the populace of Crotona, Sybaris, and other towns were excited against the Pythagorean clubs, and they were suppressed. Pythagoras himself is believed to have died at Metapontum, where his tomb was shown in the time of Cicero. It is probable that he never actually wrote anything, but that his teaching was oral.

Strabo was a native of Amasia in Pontus, living during the reigns of Augustus and Tiberius. He was born about B.C. 54. He was several years at Rome, and travelled in Egypt, and probably over the greater part of the Roman Empire. The work of Strabo is the most important geographical work that has come to us from antiquity. He fully discusses the systems of his predecessors, Eratosthenes and Hipparchus, Posidonius and Polybius, and then gives an outline of his own views. He adopts the doctrine of the sphericity of the earth, of the equator and ecliptic, and of the five zones, and the measurement of the circumference of the earth according to Eratosthenes. Considering the error caused by ignoring the curvature of the sphere to be unimportant, Strabo represented the world, on his map, as a plane surface, and the meridians and parallels of latitude as straight lines crossing each other at right angles. His first two books are introductory, and in the third he commences a particular description of the different countries. He devotes eight books to Europe, six to Asia, and one to Egypt and Ethiopia.

Thales, the Ionian philosopher, was born at Miletus, contemporary of Crœsus, 560 B.C.; and lived to a great age, but he left behind him nothing in a written form.

Thales (or Thaletas), the lyric poet and musician, was a native of Gortyna in Crete, who settled in Sparta, where he became the head of a new school of music in about 660 B.C. There are no remains of his poetry.

One of the latest of the Byzantine scholars and writers, who fled into Italy when the Turks took his native city of Thessalonica, in 1430 A.D. He taught Greek at Ferrara, and was employed by the Pope in translating Greek works into Latin. He died A.D. 1478. He wrote a Greek grammar, and translated some of the works of Aristotle.

Theon of Alexandria is best known as the father of Hypatia. Of himself personally there are no particulars, except that he was an astronomer and mathematician. He wrote a commentary on Aratus, and another on the *Almagest* of Ptolemy, addressed to his son Epiphanius.

The historian, native of Tauromenium in Sicily, and son of Andromachus, the tyrant of that colony. Timæus was born about 352 B.C., and died at a great age in 256 B.C. He was banished from Sicily by Agathocles, and passed his exile at Athens. He wrote a history of Sicily from the earliest times, which was severely criticised by Polybius, but commended by Cicero. Some fragments of the work of Timæus have been preserved.

M. Terentius Varro was born B.C. 116, being ten years older than Cicero, with whom he lived for a long period on terms of intimacy. Varro held a high naval command in the war against Mithridates, and continued to serve under Pompey until the battle of Pharsalia. He then submitted to Cæsar, who received him graciously, and he passed several years in retirement, at his country seat near Cumæ. He died B.C. 28, aged 88. He was a man of vast learning and the most voluminous of Roman authors. Only one of his books, and fragments of another, have been preserved, namely, *De re rustica*, an important treatise on agriculture; and a work on grammar.

Vitruvius was born about B.C. 76, and composed his work on architecture in the reign of Augustus. The ninth book treats of sundials and other instruments for measuring time.

The Athenian, born about B.C. 444. In B.C. 401 he joined the expedition of the younger Cyrus, and became acquainted with the Persian standards of measurement. On the defeat and death of Cyrus, the Greek contingent under Xenophon was left alone on the wide plain of Mesopotamia. Xenophon retreated across Armenia to Trebizond, and thence to Chrysopolis. On the death of his master Socrates, B.C. 399, Xenophon was banished from Athens, and joined Agesilaus, King of Sparta. He settled at Scillus in Elia, where he passed his time in writing and hunting. Here he probably composed the *Anabasis*. He lived 20 years at Scillus, and probably died at Corinth.

INDEX OF NAMES

OF

STARS AND CONSTELLATIONS

As given by Hues *in his "Tractatus de Globis",*

WITH REMARKS.

Some of the notes on the Arabic names have been kindly furnished by Professor Robertson Smith. His references are to *Cazwini*, *"Ajaib al-Makhlucat,"* (*Vol.* i, *ed. Wüstenfeld. Göttingen*, 1849). *Ideler's "Sternnamen"* (8*vo. Berl.*, 1809) gives a translation with notes; but, except in special cases, Mr. R. S. has referred to the Arabic text.

The authorities of *Hues* are *Alfraganus, Scaliger, Grotius, Jacob Christmannus*, the *Almagest of Ptolemy, Reinholt, Copernicus*, &c.

I, denotes stars of the first magnitude. II, second magnitude.

Alanin (Draco).—This is probably a misprint for *Atanin*, one of the many corruptions of *Tinnīn*: see *Assemani*, p. cii. (See *Rasaben*) 51

Alarnebet (Lepus).—الارنب *Al-arnab*, "the hare" (*Cazwini*, p. 39; *Dorn*, p. 60). *Alarnebet* represents the feminine form, which is not applied to the constellation 60

Al-asad (Leo).—الاسد (*Cazwini*, p. 36; *Dorn*, p. 54) 56

Alashan (Sagittarius).—Probably على السهم *Ala-l-sahm*, "the star upon the arrow." Cazwini names two such stars 58

Alatod (a star in Auriga), العتود *Al-atud*. In Spanish pronunciation *atod*. A yearling goat, or Capella. (See *Al-haisk*) 52

Al-atrah (Scorpio).—The corrupt form, *Al-atrab*, for *Al-acrab*, is noticed in *Assemani*, p. cliii. *t* and *c* being very like in many manuscripts and early prints, this was an easy corruption; *h* for *b* is a further corruption. (See *Al-acrah*) 58

Al-batina (Crater).—A corruption for الباطية *Al-bātiya*, "the cup" (*Cazwini*, p. 40; *Dorn*, p. 61). This is an error due to the diacritical points. (See *Elkis*) 61

Al-cheleb al-akbar (Canis).—الكلب الاكبر *Al-kalb al-akbar*, "the greater dog" (*Cazwini*, p. 39; *Dorn*, p. 60) 60

Al-cheleb-al asgar.—الكلب الاصغر *Al-kalb-al-asghar* (Anti-canis), "lesser dog" (*Dorn*, p. 61); Cazwini (p. 39) has الكلب المتقدم *Al-kalb-al-mutacaddim*, "the preceding dog," a rendering of Προκύων ... 60

Al-cheti-hala-rechabatch (Hercules).—الجاثي على ركبتيه *Al-jāthī-alā-rukbataihi*, "the kneeler on his knees." *Engonasin* (Ἐν γόνασιν) or *Nixus in genubus*. (*Cazwini*, p. 32; *Dorn*, p. 45) *tch* is a corruption for *teh* 52

Al-cusu (Sagittarius).—القوس *Al-caus*, "the bow." The final *u* is the Arabic nominative termination. It is also called الرامى, "the archer" (*Cazwini*, p. 37; *Dorn*, p. 56) 58

Al-delphin (Delphinus) 54

Aldebaran (α Tauri) I.—The Bull's eye. "*Palilicium*" of the Latins. A bright red star, at one end of the V formed by the Hyades, and one of the stars whose lunar distances are given in the Nautical Almanac. *Aldabarān* also means the Hyades as a whole, الدبران *Al-dabaran* from *Dabara*, "to follow." This is an old Arabic name for the Hyades not borrowed from the Greek, and its meaning is disputed among the Arabs. It probably designates the Hyades as following the Pleiades. See *Krehl, Religion der Araber* (1863), p. 10. (*Cazwini*, p. 35) 55

Alderaimin.—A star in Cepheus.—الذرع الايمن *Al-dhir'-al-aiman*, "the right fore arm," the article *al* being omitted before the adjective incorrectly. *Cazwini* (p. 31) describes α *Ophei* as on the left shoulder, and also speaks of stars on the fore arms ... 51

Aldigaga (Cygnus).—الدجاجة *Al-dajaja*, "the hen," and الطائر *Al-tair*, "the bird," are names of Cygnus (*Cazwini*, p. 32; *Dorn*, p. 46) 53

Alfaras-alatham (Pegasus).—الفرس الاعظم *Alfaras-al-a'zam,* "the greater horse," which is clearly what is meant here, as ظ in the more classical pronunciation is not a palatal *z* but a palatal *th.* (*Cazwini,* p. 34 ; *Dorn,* p. 50.) (See *Al-menkeh*) 54

Alfard (α Hydræ).—الفرد- *Al-fard,* "the isolated one." (*Cazwini,* p. 40 ; *Dorn,* p. 25) 61

Alfecca.—(See *Corona Borealis.*)

Alferkathan.—(Stars in Ursa Minor) الفرقدان *Al-farcadān,* "the two calves" (β and γ). The brightest of the four stars that make a square in Ursa Minor (*Cazwini,* p. 29 ; *Dorn,* p. 43) ... 50

Algedi.—(Arabic name for Capricornus) الجدي *Al-jadi,* "the kid." (*Cazwini,* p. 37; *Dorn,* p. 56) 58

Algenib (γ Pegasi).—الجنب *Al-janb,* "the side." Various constellations have stars so named, especially α Persei (*Ideler,* p. 116).

Algeuze.—An Arabic name applied to Gemini, and also to Orion. "Of the stars in Gemini that which is their head is called *Ras-algeuze.* Now *Geuze* signifies a walnut, and perhaps they allude herein to the Latin word *Ingula,* which name Festus calleth Orion, because he is greater than any of the other constellations as a *walnut* is bigger than any other kind of nut" 59

The meaning of الجوزاء *Al jauzā* is not clear, and the constellation is also called التوامان *Al-tauaman,* "the twins," after the Greek (*Cazwini,* p. 36; *Dorn,* p. 53) جوز *Jauz* means "a nut," but this word can hardly have anything to do with it. *Al-jauza* is a name also given to Orion (*Cazwini,* p. 38; *Dorn,* p. 59) ... 59

Al-gibbar (Orion).—الجبار *Al-jabbar,* "the hero." (*Cazwini,* p. 38; *Dorn,* p. 59) 59

Al-gol (β Persei).—الغول *Al-ghul,* "the ghoul": i.e. Medusa, the monster. (*Cazwini,* p. 33; *Dorn,* p. 47). (See *Chamil-ras-algol*) 53

Algomeiza (Procyon).—الغميصاء *Al-ghumaiça,* "the little watery-eyed one" (fem.), because she weeps for her separation from her brother, *Canopus.* (See the myth in *Ideler,* p. 245) 60

Algorab (Corvus).—الغراب *Al-ghorab,* "the raven." (*Cazwini,* p. 41; *Dorn,* p. 61) 61

Al-haisk (Capella).—العيوق *Al-'ayyūc* (*Cazwini,* p. 33; *Dorn,* p. 25). The European forms are 'Αϊούκ, *Alhajoc,* &c. (See *Ideler,* p. 92.) The Arabic lexicographers (*Lane,* p. 2199) take the name to mean "the impeder" of the Hyades from meeting the Pleiades; but this is very questionable. (See *Alitod*) 53

Al-hakkah (Aquila).—العقاب *Al-'ucāb,* "the eagle." Here again *ayin* is represented by *h:* and the final *h* should be *b* (*Cazwini,* p. 33; *Dorn,* p. 50). (See *Altair*) 54

Al-hamel (Aries).—الحمل Al hamal, "the young ram" (*Cazwini,* p. 35; *Dorn,* p. 52). The name of the constellation transferred to a star 55

Al-hasa (Serpens).—An error for *Al-haia* ; see the next name ... 54

Al-hava (Serpentarius).—Cazwini (p. 33) puts together, as one constellation, the serpent charmer and the little serpent, والحوية الحوّاء *Alhawwā' wal-huwayya*. A various reading for *Huwayya* is *Hayya* (*Dorn*, p. 49) 54

Al hava (Bootes).—*Al-'awwā'* العوّاء "the howler." (*Cazwini*, p. 32; *Dorn*, p. 45. See *Bootes*) 51

Aliemalija (Sirius).—Corruption of *Al-yamaniya*. (See *Alsahare*.)

Alioth (ε Ursæ Majoris).—The first star in the tail of the Great Bear. The name is from the Alphonsine Tables, and Cazwini (p. 30) calls the same star الجون *Al-jaun*. *Alya(t)* would be the buttock. 50

Almara-almasulsela (Andromeda).—(See *Alamac*.)

Almenkeh (γ Pegasi).—المنكب *Al-mankib*, "the shoulder"; or fully *Mankib al-faras*, "the horse's shoulder." (*Cazwini*, p. 34) ... 54

Al-mugamra (Ara).—المجمرة *al-mijmara*, "the censer," θυμιατήριον (*Cazwini*, p. 41; *Dorn*, p. 60) 62

Al-mutaleh (Triangulum).—المثلث *Al-muthallath*. (*Cazw.* p. 41; *Dorn*, p. 60) 55

Al-nahar.—(See *Achernar*.)

Alphard.—(See *Alfard*.)

Alphecca.—(See *Corona Borealis*).

Alpheratz (α Andromedæ).—Dorn (p. 51) gives α Andromeda as راس المسلسلة *Ras-al-musalsala*, "the head of the woman in chains." *Ratz* should be *Ras*. *Phe* may be *Fi* (في) the preposition *in*. If so, the article is incorrectly prefixed to an abbreviation of a descriptive phrase 55

Al-redaf (Cepheus).—According to Ideler (p. 297) this is a mistake. (See *Deneb-al-digaga*) 51

Al-rucaba (Ursa Minor).—The Pole Star of the Alphonsine Tables. It had not been traced to an Arabic name when Ideler wrote, and his conjectures do not carry us further (p. 15). It may be الركاب *Al-rekab*, properly "the stirrup"; which is also "point d'appui," and gives in Spanish *arrocaba*, the perpendicular beam on which a roof rests, a suitable metaphor for the Pole Star 50

Al-sahare al semalija (Procyon).—*Al-semalija* is plainly الشماليه "northern." *Al-sahare* is *Al-shi'ra*, a name applied both to Sirius and Procyon. They are distinguished as اليمانيه and الشاميه *Al-yamaniya* and *Al-shamiya*, or Yemenite and Syrian—Southern and Northern 60

Alsahare Aliemalija (Sirius).—For *al-shi'ra al-yamaniya* 60

Alsamech-alramech (Bootes).—Properly Arcturus or α Bootis, which the Arabs call *Al-simāk al-rāmih* السماك الرامح "the prop that carries a spear," to distinguish it from the other *simāk*, Spica Virginis. The spear itself (*rumh*) is η Bootis 51

Al sartan.—السرطان *Al-saratan*, "the crab" (*Cazwini*, p. 36; *Dorn*, p. 53). (See *Cancer*) 56

Al semcha.—السمكة *Al-samaka*, "the fish" 58

Al-sephina.—(See *Sephina*).

Al-soham (Sagitta).—السهم *Al-sahm*, "the arrow." (*Cazwini*, p. 33; *Dorn*, p. 50). (See *Istusi*)... 54

Al-subah (Fera).—السبع *Al-sabu'*. (See *Asida*) 61

Al-sugahr (Hydra).—الشجاع *Al-shujā'*, "the Serpent." (*Cazwini*, p. 40; *Dorn*, p. 61). The final *hr* seems to be an attempt to represent the guttural *'ayn* 61

Altair (α Aquilæ).—الطائر *Altair*, fully *al-nasr al-tāir*, "the flying vulture." Cygnus I.—A green star whose lunar distances are given in the Nautical Almanac. The name is also applied to Cygnus. (See *Alhakhah*) 54

Al-tor (Taurus).—الثور *Al-thaur*. (*Cazwini*, p. 35; *Dorn*, p. 53) ... 55

Al-vakah (Vega).—الواقع *Al-wāci'*; fuller النسر الواقع *Al nasr al wāci'*, "the falling vulture." (*Cazwini*, p. 32; *Dorn*, p. 46). (See *Lyra*, *Vega*, *Schaliaf*) 52

Anchenetenar, or *Angetenar* (a star in Eridanus).—This, according to Ideler (p. 234) corresponds, as Scaliger observed, with Ptolemy's ἐπιστροφή, or "turn of the river." The longer form seems to be due to a conjecture of Scaliger. The shorter is the Alphonsine form. This is clearly انقطاع النهر *Incita' al-nahr* (*annahr*), "the place where the river stops short and turns." Cazwini (p. 38) uses this word, and says:—"then it stops and passes, etc." 60

Andromeda, the constellation between Perseus and Pegasus, and south of Cassiopeia. *Alanac* is γ Andromedæ, *Mirach* or *Mizar* is β Andromedæ 55

Antares (α Scorpii) I. to II.—A red star whose lunar distances are given in the Nautical Almanac. *Antares* forms a right-angled triangle with *Spica* and *Arcturus*, the right angle at *Spica*.

Anticanis.—The Latin name for *Procyon* 60

Antinous.—(See *Aquila.*)

Aphelon (Castor) is a deformation, through the Arabic, of Ptolemy's Apollo. (See *Ideler*, p. 151) 56

Apollo.—A name given to *Castor*, by Ptolemy 56

Aquarius.—Constellation of the Waterman. In Arabic it is *Al-dalw*, which means "a bucket to draw water," or *al-dālī*, *al-sākib*, "the water-drawer." One star is called Seat, i.e. Arabic *sa'd* 58

Aquila.—Constellation of the Eagle. In Arabic *Al-'ucab*. There are three stars in a line, of which *Altair* is the centre. The ancients reckoned nine stars, besides six of small magnitude, which the Emperor Hadrian caused to be called *Antinous*, in memory of the Bithynian youth who sacrificed his life for his imperial master. (See *Al-hakkah*, *Altair*) 53

Ara, or *Thuribulum*.—A small constellation S, in Arabic *al-mijmara* (*Almugamra*). Bassus called it *Sacrarium* 61

Arcturus (α Bootis) I.—So called both by Greeks and Latins. The herdsman. Ἄρκτος, a bear, and Οὖρος, a warder. In Arabic *Al-simak al-ramih*, "the spear-bearing prop"; very corruptly

Somach-haramach (see *Al-samech al-ramech*). Theon placed it in the girdle of the herdsman—Bootes. It is a yellow star, the fourth brightest in the heavens, travelling at a rate of fifty-four miles a second. See Job ix, 9, xxxviii, 32, where, however, modern scholars do not understand Arcturus to be the star meant 51

Argo Navis.—Southern constellation of the ship *Argo*, called by the Arabs *Safina*, "ship." A dialectic name for *ship* is *Markab*, which also means *saddle*, and so is the name of α Pegasi. The Alphonsines give the name Mera to the sixth star in the constellation. It contains forty-five stars, the brightest being *Canopus*. The next is η *Argus*, which varies from the first to the sixth magnitude in forty-six years. (See *Sephina*) ... 60

Arided.—A star in Cygnus. (See *Deneb*) 53

Aries.—Constellation of the Ram. In Arabic *Al-hamal*. It has thirteen stars according to Ptolemy 55

α **Arietis**, or *Hamal*, the chief star of Aries, and formerly the point of intersection of the equator and ecliptic; now 27° short of it. II. Yet one of the stars whose lunar distances are given in the Nautical Almanac 55

Asida.—(A star in Fera.) According to Scaliger (*Ideler*, p. 279), a Turkish planisphere has الاسدة *Al-asada*, "the lioness," for the usual السبع *Al-sabu* (Fera) of Cazwini (p. 41) and Dorn (p. 62). (See *Al-subah*) 61

Asugia and *Asuia*, other forms of *Al-sagahr* or *Al-shuja*. (See *Orion.*)

Asumpha.—(A name of *Deneb*, in Leo.) Perhaps a corruption of *Al-çarfa*, which is the Arabic name of β Leonis 57

Atauri.—(See *Al-tor.*)

Auriga.—Constellation of the Waggoner. In Arabic, *Mumsik-al-a'inna* (corruptly *Memassich al hanamshat*), "He who holds the reins," and *Roha*. It contains fourteen stars, the bright one in the left shoulder being *Capella* 53

Beemim (γ Eridani).—Ideler (p. 234) has conjectures as to derivation, but they are not satisfactory 60

Bellatrix (γ Orionis).—A bright blue star in the shoulder of Orion, so named in the Alphonsine Tables 59

Benetnasch (η Ursæ Majoris).—بنات نعش *Banat na'sh*. (*Cazwini*, p. 30.) *Na'sh*, "the bier," is the four stars that form the square of the Great Bear; and the three tail stars are the daughters who follow it—the "*Filiæ Feretri*." The proper Arabic name of η Ursæ Majoris is *Al-caid*, "the Governor"; and *Banat na'sh*, or "the bier's daughters," is the name of the seven stars as a whole 51

Berenice's Hair.—Stars between *Leo* and *Ursa Major*, reckoned by Theon as belonging to *Virgo*. Conon, the mathematician of Alexandria, dedicated them to the hair of Ptolemy's wife, and the poet Callimachus celebrated the event in verse 57

Centaurus.—A constellation of thirty-seven stars, called by the same name in Arabic. Among the stars, those in the Centaur's feet now form the Southern Cross. α *Centauri* I. is the nearest of the fixed stars, being only 200,000 times the distance of the sun from the earth. It is a double star. β *Centauri* I. is equally near 61, 66

Cepheus.—A constellation of eleven stars, called *Phicares* by the Phœnicians, which is interpreted as *Flammiger*. It contains a star called by the Arabs *Al-dhir'-al-aiman* (corruptly *Alderaimin*), "the right fore-arm." (See *Alderaimin*) 51

Cetus.—Constellations of the Whale, called *El caitus* in Arabic. Of its twenty-two stars, *Menkar-el-kaitos* means "the Whale's snout," *Boten-el-kaitos*, "the Whale's belly," *Deneb-el-kaitos*, "the Whale's tail." *o Ceti* is a star which changes from first to twelfth magnitude in 331 years 59

Chamil-ras-algol.—A star in Perseus. حامل راس الغول. *Hāmil ras al-ghul*, "He that carries the Ghoul's head." (*Cazwini*, p. 33.) The name of the constellation transferred to its chief star. The star is usually called *Algol* 53

Chesil.—The Hebrew name for Canopus 61

Corona Australis, or the constellation of the Southern Crown. In Arabic, *Alachil-al-genubi* (correctly *Aliklīl-al-janūbī*). It is formed of thirteen stars forming a double wreath 62

Corona Borealis, or the Northern Crown. Arabic, *al-iklīl al-shamālī*. The bright star which seems to fasten the chaplet is called *Alpheca* (*al-fakka* الفكة), meaning *Solutio*, "untying"; also called *Munic*. The constellation has eight stars. η and γ Corona Borealis are double stars; and τ Coronæ (previously of tenth magnitude) blazed out suddenly in May, 1866 52

Corvus.—Constellation of the Crow; in Arabic, *Al-ghorab*. It has seven stars 61

Crater.—Constellation of the Cup; in Arabic, *Al-batiya*, or *Al-kas* (corruptly *Elkis*). It has seven stars 61

Crucero, or the constellation of the Southern Cross, detached from the Centaur, of which it formed the hind feet. The Elizabethan navigators corrupted "Crucero" into "*the Cruisers*." It was only known to Ptolemy as the *Centaur's feet*. α and β Crucis are I.I. 67

Cygnus or Gallina.—Constellation of the Swan or Hen, of seventeen stars. In Arabic it is called *Al-dajaja* (corruptly *Aldigaga*) and *Altair*. The chief star is *Deneb*, which see. β and δ Cygni are double stars 53

Delphinus.—The constellation of the Dolphin, ten small stars. In Arabic, *Al-delfin* 54

Deneb (α Cygni) or *Deneb-aldigaga*, ذنب الدجاجة *Dhanab-al-dajaja*, "the hen's tail." (*Cazwini*, p. 32.) Cazwini says that the bright star on the tail is called الردف *Al-ridf*, "the one who rides behind" the four fawāris or horsemen (γδεζ) (*Dorn*, p. 46).

Also *Arided*, which is a corruption of *al-ridf*. It is a green star 57

Deneb-al-asad (β Leonis).—ذنب الاسد *Dhanab-al-asad*, "lion's tail." The usual name in Arabic is الصرفة *Al-çarfa* (*Cazwini*, p. 46). *Dhanab-al-asad* is given as a synonym by *Alferghani apud Assem.* cxc. 57

Deneb-al-gedi (Star in Capricornus).—ذنب الجدي *Dhanab-al-jadi*, "kid's tail" 58

Deneb-al-kaitos (Star in Cetus).—ذنب القيطس *Dhanab-al-caitus*, "tail of the κῆτος." There are several stars at the root of the tail, according to *Cazwini* (p. 38) called النظام *Al nizam*, "the string of pearls," and one in its southern part called "the second frog," الضفدع الثاني... 59

Denebola (β Leonis).—A white star. Ola may be for *aula*, "first"... 57

Dhath-al-cursi (Cassiopeia).—ذات الكرسي *Dhāt-al-kursī*, "the woman with the chair." (*Cazwini*, p. 32 : *Dorn*, p. 46.) 53

Dobhe (α Ursæ Majoris).—Called by Dorn (p. 43) ظهر الدبّ, *Zahr-al-dubb*, "the bear's back." The last word only has been retained, and the final *e* may represent the genitive termination, or, as Dorn suggests (p. 69), it may represent the feminine *Dubba, Ursa*. Ideler (p. 23) supposes that the name of the constellation has simply been transferred to the chief star, as in other cases 50

Draco.—A constellation of thirty-one stars 51

Dub-al-akhbar (Ursa Major).—الدبّ الاكبر, *Al-dubb-al-akbar*, "the greater bear" 50

Dub-al-asgar (Ursa Minor).—الدبّ الاصغر *Al-dubb-al-asghar*, "the lesser bear" 50

Echer.—(See *Sirius.*) A corruption of *Al-shi'rā*.

El-adari (Virgo).—العذراء *Al-'adhra*, "the Virgin." (*Cazwini*, p. 36 ; *Dorn*, p. 54) 57

El-cusu (Sagittarius.)—See *Al-cusu* 58

El-delis (Aquarius).—الدلو. *Al-dalw*, "the bucket." (*Cazwini*, p. 37) 58

El-kaitos (Cetus).—القيطس. *Al-caitus*, a transcription of κῆτος (*Cazwini*, p. 31) 59

Elkis (Crater).—الكاس *Al-kas*, "the cup." Pronounced in Spain Al-kēs, which is the form in the Alphonsine Tables (*Ideler*, p. 271) for α Crateræ. (See *Al-batina.*) 59

Elgueze —(See *Al-geuze.*)

El-seiri (Sirius).—الشعري *Al-shi'ra*, Σείριος. The Greek word is itself probably a loan word, and the Arabic not merely a copy of it. It may mean "hairy." In Arabian astronomy there are two *Shiras*, Sirius and Procyon 60

Enif-alfaras (Pegasus).—Obviously انف الفرس *Anf-al-faras*, "the horse's nose." Ideler (p. 116) identifies it with فم الفرس *Fum-al-faras* of Cazwini (p. 34), "the horse's mouth," ε Pegasi ... 55

Equiculus.—The constellation of the Little Horse. In Arabic, *cit'at al-faras*, προτομὴ ἵππου, i.e. "fore part of a horse cut off." Four obscure stars 54

Eridanus.—Constellation of the River, called in Arabic, *Al-nahr*. It consists of thirty-four stars. The Arabs called one star *incitā'al-nahr*, "the turn of the river" (corruptly *Anchetenar* or *Angetenar*), and another *Beemin* (which see). *Akhir-al-nahr*, known as *Achernar* (which see), is another bright star in this constellation 59, 66

Fera.—An obscure constellation .called *Asida* in Arabic, and *Al-sabu'* (corruptly *Alsubah*). Nineteen small stars 61

Flammiger.—(See *Cepheus*).

Fomalhaut (Piscis Australis).—A bright white star I. to II. Its lunar distances are given in the Nautical Almanac. فم الحوت, *Fum-al-hut*, "mouth of the fish." (*Cazwini*, p. 41) 62

Gallina.—(See *Cygnus*).

Gemini.—Constellation of the Twins, consisting of eighteen stars. In Arabic, *Algeuze*. Some will have the twins to be *Castor* and *Pollux*, others *Apollo* and *Hercules*. With the Arabians one is called *Aphelon* and *Aellar*, the other *Abracaleus* or *Gracleus*, as Scaliger conceives 56

Gibbar.—(See *Sirius*).

Gracleus.—A name of Pollux 56

Habor.—Corrupted from *Echer*, (which see).

Hain-altor.—Bull's eye ; Ar. عين الثور *'ayn al-thaur*. (See *Aldebaran*.)

Hamel (α Arietis).—(See *Al-Hamel*.)

Has-alangue.—(See *Al-hava*). Should be *Ras-alangue?* Perhaps rather a corruption of the star called *asl-dhanab al-hayya*, "root of the serpent's tail" (*Dorn*, p. 49)... 54

Hazimath al-hacel (Spica).—Very corrupt for السماك الاعزل, *Al simak al a'zal*, "the unarmed prop": as distinguished from the spear-bearing *Simak*, or *Arcturus* (*Al-simak al-ramih*) (*Cazwini*, p. 47), or because it does not bring wind or cold. (See *Lane*, p. 1430) 57

Hædi.—"The Kids." (See *Capella*.)

Hercules.—A constellation of eight stars. In Arabic, *Al jathi alà rukbataihi* (corruptly *Alcheti hala rechabatah*), "the kneeler on his knees." The Latins called it *Nisus* or *Nixus*. The star in the head is *Ras-al-jathi* (*Rasacheti*), "the head of the kneeler"; not *Rasaben*, as the Alphonsines corruptly have it. Another star is *Marfic* (corruptly *Marsic*) or "the elbow," another *Miçam* (corruptly *Maasim* or *Mazim*) "the wrist": corresponding to κ and ο Herculis. The sun is now approaching Hercules at a rate of four miles a second. ζ Herculis is a double star 52

Hyades (In Taurus).—(See *Aldebaran*.) Thales Milesius says there are two, Euripides three, Achæus four, Hippias seven 56

Hydra.—In Arabic *Al-shuja'* (*Alsagahr*) and *Asugia*, "the serpent." The constellation consists of twenty-five stars, one of them (α) called by the Alphonsines *Alfort*, i.e. *al-fard*, "the isolated." The Egyptians called it *Nilus* 61

Ingula.—(See *Al-geuze*).

Marsic or **Mazim** (a star in Hercules).—Ideler (p. 65) explains these two names from the Alphonsine Tables correctly. *Marsic* is an error for *Marfic* مرفق "elbow": and *Masym* is for *mi'çam* معصم, "wrist." They correspond to κ and ο Herculis 52

Megrez (γ Ursæ Majoris).—(See *Phegda.*) Or δ is *Megrez*, and γ *Phachd*, or "the thigh."

Mellef (a star in Cancer).—المعلف *Al-mi'laf*, "the crib or manger." "*Praesepe*" (*Cazwini*, p. 36). Cazwini says that this is the name in the Almagest, a translation, therefore, of φάτνη ... 56

Memassich-al-hanamshat (Auriga).—Auriga is called ممسك الاعنة *Mumsik-al-a'inna*, "he who holds the reins," Ἡνίοχος: and also *Mumsik-al-inan* العنان, "he who holds the rein." (*Dorn*, p. 48.) *Mumassik* will mean the same. The first part of *Hanamshat* is clearly *'inan* with *h* for *ayn*. It is possible that *shat* may be سائطا "wielding the whip," or *sayyāt* "the whipper." Another name is *Roha* 53

Menkar (a star in Cetus).—منخر *Mankhar* or *Minkhir*, "the nostril." According to Ideler (p. 210) it is λ Ceti 59

Merak (β Ursæ Majoris).—المراقّ *Al maracc*, "the loins." (*Dorn*, p. 43.)

Mirach or **Mizar** (a star in Cassiopeia).—مئزر *Mizar*, "drawers" or "waist cloth." Cazwini (p. 34), and Çufi *apud* Dorn (p. 58) speak of the *Mizar* or "waist cloth" of Andromeda. The same part of the body can equally be called *Maracc*, "the loins," المراقّ. 53

Mizan aliemin (Libra).—الميزان *Al mizan*, "the balance." The second word may be اليمين *Al yamin*, "the right hand," so that the name would properly denote the southern scale, or is it *al-aiman*, "the lucky"? 57

Mirzar (ζ Ursæ Majoris).—Ideler (p. 24) writes *Mizar*, and supposes that, as in the case of Andromeda, it was originally *Merak*, or β Ursæ Majoris, and has changed its place. But this involves two mistakes, for a bear would not have a waist cloth. Cazwini (p. 30) and Dorn (p. 43) call this star "the goat"—*Al 'anac* العناق. A synonym would be معزى *Mi'za*. It seems very likely that this is the true origin of the word, the *r* being added by false analogy. (See *Phegda*) 50

Moselek.—(See *Schomlek*).

Munic.—(See *Corona Borealis*).

Mutlathan.—(See *Almutaleh*).

Nesses.—Corruption of Nisus, by Vitruvius.

Nilus.—(See *Hydra*).

Nisus or **Nixus.**—(See *Hercules*).

Orion.—Sometimes called *Asugia* (*Al-shujā'*, "the valiant man," the same Arabic word that also means "water-snake or hydra";) (but is there any proof that this name really means Orion?) and sometimes *Al-geuze* by the Arabs: also *Al-gibbar*, "the hero."

The constellation contains thirty-eight stars. *Betelgueze* on the right, *Bellatrix* on the left shoulder, *Rigel* the foot, and three small stars form the belt,—δ, ϵ and ζ Orionis. (See *Job* xxviii, 31; and *Amos* v, 8) 59

Palilicium.—(See *Aldebaran*).

Pegasus.—A constellation of ten stars, called *Alfaras-alathan*, "the Great Horse," in Arabic. *Algenib* is γ Pegasi. The star on the right shoulder is *Al menkeh*, also called *Seat-alfaras;* another is *Enif-alfaras*, "the horse's nose." *Markab* is α Pegasi. β Pegasi shows, by its spectrum, that it contains hydrogen, sodium, magnesium, and perhaps barium 54

Perseus.—A constellation of twenty-six stars, in Arabic called *Chamil ras algol*, "He that carries the head of Medusa." The star over the left hand is called *Ras algol.* The Alphonsines named one of the stars *Algenib*, meaning "the side." β Persei is of second magnitude for two-and-a-half days, then suddenly falls to fourth magnitude in three hours; returns in the same time ... 53

Phegda (δ Ursæ Majoris).—This is evidently فخذ *Fakhidh*, "the thigh." The thigh is given by the authorities in Ideler (p. 22) as γ Ursæ Majoris. But Dorn (p. 43) calls γ the left thigh, and δ might very well be taken for the right thigh. Ideler, from his Eastern authorities, calls it مغرز *Maghriz*, "root of the tail," *Megrez* of the maps. If the right thigh is placed at δ and the buttock at ϵ (see *Alioth*) ζ will be the real root of the tail, and *Mirzar* or *Mizar* may be a corruption of *Maghriz* : *r* for the rolled *gh* is not unnatural.

Phicares.—(See *Cepheus*).

Piscis.—A zodiacal constellation of thirty-four obscure stars, called *Alsemcha* in Arabic 58

Piscis Australis.—A constellation of twelve stars according to Ptolemy, called in Arabic *Al-hut-algenubi.* The bright star in the fish's mouth is *Fomalhut* 62

Pleiades.—A group of six or seven small stars on the back of Taurus, increasing to sixty or seventy under the telescope. The Latins called them *Vergiliæ*, the Arabs *Al-thurayyā*. Pliny and Vitruvius place them in the tail of the Bull, and Hipparchus on the left foot of Perseus 56

Polaris—or the Pole Star, is the last star in the tail of the Little Bear. It was anciently called the Dog, and was known as the *Cynosure* (κύνος, gen. of κύων a dog, and οὐρά tail). The Phœnicians always steered by Polaris as Aratus affirms, while the Greeks used the Great Bear. It is less than 1° 30′ from the Pole, will approach to within 30′, and then recede. (See *Al-rucaba.*) Polaris is a white star. Its distance from the Pole in the time of Hipparchus 50

Pollux (Star in Gemini) I. to II.—Called *Hercules* by some, *Abraceleus* for *Bracleus* by the Arabs, as Scaliger conceives. The lunar distances of Pollux are given in the Nautical Almanac. Pollux contains iron, hydrogen, sodium, and magnesium 56

Præsepe.—(See *Cancer-Mellef*).

Procyon.—In the constellation of *Anticanis* or the Lesser Dog I. It contains two stars. (See *Al-cheleb al asgar* and *Al-sahare* and *Algomeiza*). Procyon is a blue star 60

Rasaben.—(See *Rastaban*).

Rasalangue.—(See *Al-hava*).

Rasacheti (Star in Hercules).—راس الجاثى, *Ras al jathi*, "head of the kneeler." α Herculis. (See *Al-cheti*) 52

Rastaban (γ Draconis).—A star in the Dragon's head. *Ras-al-tinnin*, "the Dragon's head," is the usual name. But *taban* is plainly ثُعْبَان *Thu'ban*, one of the many Arabic words for a serpent, which is said to be the modern use for "Draco." *Rastaban* became *Rasaben*, and then the constellation as a whole (*Ras* being dropped) became *Aben*. This star is of historical interest, as a change in its polar distance attracted Bradley's attention in 1728, and led to the discovery of aberration. (See *Aben*) 51

Regulus (α Leonis) I. to II.—The bright white star in the constellation of Leo, called in Arabic *Calb-al-asad*, "the lion's heart." In Greek Βăσῐλικός, in Latin *Regulus*, because, says Proclus, those who are born under this star have a kingly nativity. The lunar distances of Regulus are given in the Nautical Almanac 57

Rigel I.—The bright blue star in the foot of Orion. In Arabic *Rigel-algeuze*, and *Rigel-al-gibbar*, رجل *Rijl*, "foot." (*Cazwini*, p. 38) ... 59

Roha.—(See *Memassich-alhanamshat* or *Auriga*).

Saclateni—of the Alphonsines, *Sodateni* of Scaliger, the Latin name for the *Hœdi* or kids, attending on Capella 54

Sacrarium.—(See *Ara*).

Saltatores.—(See *Ursa Minor*).

Sagitta or **Telum.**—A small constellation of five stars, also called *Istusi*, which word Grotius thinks is derived from the Greek οἰστός, an arrow. In Arabic *Al-sahm* 54

Sagittarius.—Constellation of the Archer, containing thirty-one stars. In Arabic *Al-caus*, *El-cusu*, "a bow" 58

Schaliaf (Lyra).—Misread, by confusing ف and ق for *Shalyac* شلياق (*Cazwini*, p. 32). It is not an Arabic word, and the first syllable points to a derivation from the Greek χέλυς, *sh* in Arabic often being used for the Greek χ. (See *Lyra*, *Vega*, *Al-vakah*) 52

Scheder (α Cassiopeiæ).—So given by the Alphonsines, but Scaliger has *Seder*, meaning a breast, صدر *Sadr*, "breast" 53

Schomlek (a star in Scorpio).—A gross corruption for شولة *Shaula*. (See *Leschat*.) Scaliger would have *Moselek* 58

Scorpio.—A constellation of twenty-one stars; in Arabic *Al-acrao*, corrupted into *Alacrah*. The star in the heart is *Calb-al-acrab*, "the sting," *Leschat*. Scaliger thinks *Schomlek* should be *Moselek* 58

Seat (a star in Aquarius).—Correctly سعد *S'ad*, "luck." Various stars are so called, with a defining word, and there are three of

Thuribulum.—(See *Ara.*)

Triangle.—An obscure constellation of four stars, called in Arabic *Almutaleh* or *Mutlathan*, which means "triangle" 55

Ursa Major.—The constellation of the *Great Bear* or *Charles's Wain;* in Arabic, *Dub-al-akhbar.* The first star in the back is *Dubhe* (α Ursæ Majoris) κατ' ἐξοχήν. *Dubhe* and *Merah* (α and β) are the pointers. *Megrez* and *Phegda* (γ and δ) are the two stars which complete the trapezium. *Alioth* (ε) is the first star in the tail, *Mirzah* (ζ) the second, and *Benetnasch* (η) the last. The constellation was first invented by Naplius, according to Theon, in all twenty-four stars. Both Bears, according to Aratus, are called ἅμαξα, "a chariot." The Arabs call the seven stars *Banatnash,* or "Filiæ Feretri," daughters of the bier. The Greeks, in navigating, were guided by the Great Bear rather than by Polaris. The Greeks called the Great Bear ἑλίκη 50, 51

Ursa Minor.—The constellation of the *Little Bear;* in Arabic, *Dub-al-asgar* and *Al-rucaba.* *Polaris* is the last star in the tail; so called because it is nearest of any to the Pole. There are two other stars in the tail, called by the Greeks χορευταί, that is, *Saltatores* or dancers. The Arabs call the two stars in the fore part of the body *Alferkathan.* This constellation of seven stars is said to have been invented by Thales, who called it the *Dog,* as Theon (upon Aratus) affirms; *Cynosure* was the dog's tail, with the pole star. (See *Al-rucaba, Al-ferkathan*) 50

Vega (α Lyræ) I.—A bright green star, which will hereafter, in 12,000 years, become the pole star, approaching within 5° of the Pole. So named by the Alphonsines. (See *Alvakah.*) Vega contains hydrogen, iron, sodium, and magnesium 52

Vergiliæ.—(See *Taurus* and *Pleiades.*)

Via lactea, or Milky Way 65

Virgo.—Constellation of the Virgin; in Arabic, *Eladari,* but more frequently called *Sunbula,* which signifies "an ear of corn." It contains twenty-six stars, the brightest being *Spica* 57

Zubeneschi-mali and *Zuben-algenubi,* in the constellation of Libra, الزبانا or الزبانى *Al-zubānā,* are α and β Libræ. (*Cazwini,* p. 47.) According to the lexicographers this is a singular form, the dual being *Al-zubanayan,* but popularly *Al-zubānān;* whence the common *Azubenen.* Thus the name of one of the two stars would be *Al-zuban,* pronounced in Spain *Al-zubēn.* And thus the two are *Al-zubān al-shamali,* the northern, and *al-janubi,* the southern الشمالى الجنوبى. The word appears to be of Persian origin, and to mean anything tongue-shaped. (See *Libra.*)

INDEX

OF

PLACES MENTIONED BY ROBERT HUES IN THE "TRACTATUS DE GLOBIS".

INDEX TO SUBJECTS.

SAILING DIRECTIONS

FOR THE

CIRCUMNAVIGATION OF ENGLAND

AND FOR

A VOYAGE TO THE STRAITS OF GIBRALTAR.

(From a 15th Century MS.)

SAILING DIRECTIONS

FOR THE

CIRCUMNAVIGATION OF ENGLAND,

AND FOR A VOYAGE TO

THE STRAITS OF GIBRALTAR.

(FROM A 15TH CENTURY MS.)

EDITED, WITH AN ACCOUNT OF THE MS.,

BY JAMES GAIRDNER,

OF THE PUBLIC RECORD OFFICE;

And a Glossary

BY E. DELMAR MORGAN,

HON. SEC. HAKLUYT SOCIETY.

LONDON:

PRINTED FOR THE HAKLUYT SOCIETY,

4, LINCOLN'S INN FIELDS, W.C.

M.DCCC.LXXXIX.

CONTENTS.

CIRCUMNAVIGATION OF ENGLAND.

ACCOUNT OF THE MS.

Among the Lansdowne MSS. in the British Museum is a folio volume, No. 285 of that collection, "the greatest part of which", as we are informed in the catalogue, "formerly belonged to Sir John Paston, Knight, in the reign of Edward IV, and was copied for him by one William Ebesham, a scribe by profession". It consists of a number of short tracts, mostly relating to pageants, coronations, challenges, tournaments, and feats of arms. Chivalry was the great study and amusement of the age, and Sir John Paston shared in the general feeling. There are, however, two treatises of more considerable length; the one a translation of Vegetius' *De Re Militari*, the other Lydgate's poetical translation of Aristotle's *De Regimine Principum.* There is also the tract here printed for the first time, containing directions for the circumnavigation of England.

That the MS., or the greatest part of it, did, as the catalogue says, formerly belong to Sir John Paston, appears at first sight to rest on indisputable evidence. There can be no doubt about the antiquity of the handwriting, and that the greater part of the contents was written by William Ebesham, Sir John Paston's transcriber, of whose signature Sir John Fenn has given us a facsimile from one of the lost Paston letters.[1] Moreover, we have in that correspondence William Ebesham's bill, delivered to Sir John Paston, for transcribing a MS. of precisely similar character; and, further, we have the description of just such a MS. in a catalogue of Sir John Paston's books. What stronger evidence could reasonably be expected? Taking even the last point alone, how very exact is the following description in an inventory of books written either by Sir John Paston or by his brother after his death:—

"Memorandum, my boke off knyghthod and the man[er] off makyng off knights, off justs, off Tor[neaments,] ffyghtyng in lystys, paces holden by so[ldiers,] and chalenges, statuts off Weer, and *De Regim*[*ine Principum.*]
Valet, . . ."[2]

Nothing could well tally more closely than this with the contents of the Lansdowne volume. And, as if to close the door to any other surmise, the catalogue states that three of the smaller articles in this volume are in Sir John Paston's handwriting, whose signature is attached to one of them at folio 42.

Nevertheless, the case is not quite so complete as this seemingly invincible evidence would make it. In the first place, the statement about Sir John Paston's hand and signature is wrong. The name of Sir John Paston does indeed occur at the end of one

[1] Fenn's *Original Letters*, vol. ii, plate v, No. 20.

[2] *Paston Letters* (new edition), iii, 301.

article, but it is certainly not a signature, nor is there any handwriting in the volume which bears the least resemblance either to that of the Sir John Paston who died in Edward IV's time, or to that of his brother John, who was knighted after him, in the days of Henry VII. This the compiler of the catalogue would probably have discovered if he had been able to examine any of the original Paston letters for comparison. But in those days they were not accessible, and his surmise, though natural, turns out to be unfounded. All that can be said is that an article written in a different hand from Ebesham's is subscribed with the words, "Quod Sir Jhon Paston", whatever that subscription may imply.

Then, as regards the notices supposed to refer to this volume in the Paston Letters. There is no doubt whatever that several of the treatises contained in this volume were actually transcribed for the first Sir John Paston by the hand of William Ebesham; for, among the documents printed by Fenn, is Ebesham's own bill for transcribing these treatises among other things.[1] The items of this account are a somewhat singular mixture of law and literature;—first, "A litill booke of Pheesyk", for transcribing which the charge is twenty pence; then, some privy seals and depositions of witnesses, some on parchment and some on paper. But the entries which concern our purpose are those at the end of the document, which are as follows:

[1] See *Paston Letters*, vol. ii, No. 596 (new edition), or in Fenn's edition, vol. ii, Letter xxiv.

"Item, as to the Grete Booke—First, for wrytyng of the Coronacion, and other tretys of Knyghthode, in that quaire which conteyneth a xiij levis and more, ij$^{d.}$ a lef . . . . ij*s.* ij*d.*

"Item for the tretys of Werre in iiij books, which conteyneth lx levis after ij$^{d.}$ a leaff . . . x*s.*

"Item, for *Othea* pistill, which conteyneth xliij leves . . . . vij*s.* ij*d.*

"Item, for the Chalengs, and the Acts of Armes, which is xxviij$^{ti.}$ lefs . . . . iiij*s.* viij*d.*

"Item for *De Regimine Principum,* which conteyneth xlv$^{ti.}$ leves, after a peny a leef, which is right wele worth . . . . iij*s.* ix*d.*

"Item, for Rubrissheyng of all the booke . . iij*s.* iiij*d.*"

If "the Grete Booke" comprised all the articles mentioned in these different items[1] it certainly bore a wonderful resemblance to the Lansdowne volume, and much certainly might be said in favour of their identity; but there are difficulties in the way. Of these five consecutive items four do indeed correspond in character and substance with different portions of the volume, and in two of these cases the number of leaves which the tract actually occupies is precisely what is stated in the account. But it is

[1] There can be little doubt that this is implied; for the writer acknowledges he had been paid some of the items in his bill, and it is the "Grete Booke" for which he specially demands payment in the accompanying letter. Moreover, though his arithmetic is a little unsatisfactory, it appears that the sum remaining due to him was 41*s.* 1*d.*, of which, as we may infer from the note added to the amount, the principal part was for the "Grete Booke". Indeed, his charge for this, if I do not misread, is 27 shillings ("unde pro magno libro scripto, xxvij *(sic)*, cum diu' chal"; which last expression evidently means *cum diversis chalengiis*, not *cum diurnali challengiorum*, as I suggested in a foot-note in the *Paston Letters*).

singular, to say the least, that the order in which they stand in the MS. is different from that of the account. Moreover, the "tretys of Werre", in four books, covers not sixty leaves, but only fifty-three, and a quarter of a page more. Also the treatise *De Regimine Principum* occupies, not forty-five leaves, but only forty-four; and, further, there is nothing in the volume corresponding to "*Othea* pistill". That expression we know denotes a treatise upon Wisdom. The Greek invocation 'Ω θεὰ had been converted by the ignorance of the Middle Ages into a proper name, and we meet with this divinity addressed in one poem :—

"Othea, of prudence named godesse."[1]

But nothing like a treatise on Wisdom filling forty-three leaves of paper can be found in the Lansdowne volume; and, if this be altogether a separate treatise, how comes it to be thus inserted in the account among items which are distinctly portions of the "Grete Booke"? Nor do our difficulties end even here; for, surely, in his charges for transcribing the book Ebesham might have been expected to follow the order of the contents of the book itself. But, after "the Coronacion and other tretys of knighthode" which undoubtedly stand first in the volume, he goes on to notice "the tretys of Werre", which begins at f. 83,[2] before the Challenges and Acts of Arms

[1] *Third Report of the Historical MSS. Commission*, p. 188.

[2] I follow the original contemporary foliation, in Roman figures, which, it is to be regretted, any one ever attempted to supersede, though it might have been supplemented by modern figures where it is discontinued.

which form the second portion of the volume beginning at f. 14. And it must not be supposed that the contents have been bound up in later times in a different order from that in which they originally stood; for the leaves are numbered in a contemporary hand from leaf 1 to 86, and though for a few pages this foliation (strangely enough) is dropped, it is resumed quite correctly at folio 100, and goes on to 144 in the same original hand, after which it is continued in antique numerals, but in a more modern hand, as far as f. 155. If, therefore, this MS. be the "Grete Book", referred to by Ebesham in his account, it is certain that he cited the contents in a wrong order, made two slips as to the number of leaves each article occupied, and entered one charge for a treatise not in the book at all among those which really do belong to it.

Such an amount of error is scarcely conceivable in a bill so methodically drawn up, even though the writer was, as he himself says, at the time driven to live in sanctuary to escape his creditors. Yet, it is not altogether impossible, Ebesham may have written out the items only from memory and put things down in a wrong order. There is, however, another theory which, I am inclined to think, will account more satisfactorily for these discrepancies. A professional transcriber, no doubt, copied and recopied the same treatises often for various customers, and though the contents are very much the same there is nothing positively to show that the Lansdowne volume was Sir John Paston's copy of the "Grete

Booke" at all. On the contrary, the expression, "Quod Sir Jhon Paston", strikes me, upon reflection, as if it might fairly imply that the article to which it is appended was an extract from one of Sir John Paston's MSS., taken by his permission, and that these words were added to verify the authority.

What is known of the history of the volume seems rather to suggest that it was compiled for the use of an officer of arms. The earliest owner of it, whose name we can positively ascertain, was Sir Thomas Wriothesley, Garter, in whose handwriting, as Sir F. Madden believed, some of the later entries are written, and whose initials, "T. Wr.", may be seen on the first leaf. Now, Sir Thomas lived within a generation certainly of the first owner of all, for he died in 1534; and, after his day it passed through the hands of a long line of heralds, bequeathed apparently from one to another as an official heirloom. As stated in the catalogue, "it appears to have been in the possession of Sir Gilbert Dethick, and his son Sir William Dethick, Garter King-at-Arms, and afterwards became the property of Richard Saint George, Clarencieux". The notices of its further descent are here a little interrupted. Richard, or rather Sir Richard Saint George, the friend of Camden, Spelman, and Sir Robert Cotton, was the father of a line of heralds extending to the days of George I, and we may reasonably believe that it must have remained in his family at least for a generation or two. But the next person, in whose possession we find evidence that it existed, is Sir Joseph

Jekyll, Master of the Rolls in the reign of George I and II. From him, however, it again passed into the possession of a herald, the industrious antiquary Oldys, who made considerable use of it in his article on Caxton in the *Biographia Britannica*. After his death it was, doubtless, acquired by the first marquis of Lansdowne, and thus became a portion of the Lansdowne library, now in the British Museum.

So much for the history of a MS., the general contents of which possess an interest for the historian and the antiquary quite apart from that of the one brief article here edited for the Hakluyt Society. It only remains to say that that article is written in the clear business-like hand of William Ebesham, and though the punctuation is defective, and the spelling, of course, not more uniform than that of the very best penmen of the age, there is not a single letter throughout which is either illegible or uncertain, except where combinations occur of the letters *m*, *n*, *u*, and *i*. These letters, as every one knows who is at all familiar with the handwritings of the period, were invariably expressed, when in the middle of a word, merely by a number of upright strokes called *minums;* no difference whatever was made in the formation of the letters *n* and *u*, and the *i*'s had no dots to distinguish them. Hence, ambiguities may occur occasionally as to the spelling of proper names, only known to us through a unique mediæval MS. like the present.

SAILING DIRECTIONS

FOR THE

CIRCUMNAVIGATION OF ENGLAND

AND FOR

A VOYAGE TO THE STRAITS OF GIBRALTAR.

BERWIK lieth south and north of Golden stonys, the Ilonde and Berwik haven lien west north west and Est South est. And fro Vamborugh to the poynt of the Ilond the cours lieth north and South. And beware of the golde stonys it folowith North north west, and quarter tide be owte fro Tilmouth to Fenyn Ilonde the cours is North northwest and South South est. And Tilmouth is tide north est and South west betwene the hedelonde and houndeclif fote the cours is northwest and southest. And it flowith west southwest and Est northest. And at Whitevies half and fro Houndeclif fote to Humbre the cours is south est and be south, northwest and by north. Fro Leyrnes to the Hedelonde the cours is north northwest and south south est, at the Hedelonde the streme settith North West and Southest, and it flowith on the londe of Holdernes northest, and quarter tide in the faire way, and at Hedelonde quarter tide and half. And yif ye go from Leirnes to the Shelde ye shall goo Est Southest for to go cleene of Resande and by South. And yif ye have an ebbe go southest and by Est. And yif ye go fro the spone to the shelde, and that the wynde be at Northwest your cours is Southest till ye be passid Welbank. And in well it flowith est and west, And there goeth half streme undir Rothir. And at the shelde it flowith on the londe West north west and half streme undir Rothir by the londe till ye come to Winterbornes. And from Wyntirburnes till ye coome to

Cukle rode it flowith on the londe northwest and quarter tide and half quarter undir Rothir. And yif ye goo fro the shelde to the Holmes, and it be in the nyght, ye shall go but xviij fadome fro the coste till the gesse that ye be past Limber[1] and Urry, and to the estermare cours till ye come to xiiij fadome. And go your cours South southest till ye be passid the Holmes, but the moost wisedome is to abide till it be day. From Kirkleholmys to Orfordnesse and the wynde be on the londe saile your way is south and by west it flowith on the londe south southest, and at the Holmys hede quarter tide, fro Orfordnesse to Orwell waynys the cours is southwest and it flowith south southest. And in Orwell haven within the weris south and north, and yif ye go oute of Orwell waynys to the Naisse ye must go south west fro the Nasse to the merkis of the spetis your cours is west south-west, and it flowith south, and by Est bring your markis to gidre that the parissh steple be owte by est. the abbey of Seint Hosies. than go your cours on the spetis south till ye come to x. ffadome or xij. than go your cours with the horse shoo south southwest, and yif it be on flode come not by in viij. ffadome, and that shall bryng you to a xj. or xij. fadome, than go your cours in to Temesse with the grene bank west southwest, and at the hors shoo it flowith south and north, and oute of Orwell waynys for to goo oute at the slade your cours is est southest, for cause of the rigge and the Rokkis, till ye com till xv. ffadome depe and for the long sande than ye may goo south southest till ye come to xvij. or xviij. fadome depe. than must ye go south a glas or two by cause of the Rokke. than goo south southwest, and seke up Tenet, and seke up vj. fadome on the brakis. than go your cours south it is your fairway, and at the Knak, in the Kentisshe See it flowith south, and at the northhede of Godewyn the streme renneth to the south south west, and it flowith from Tenet unto Wiet on both sides on the maylaunde south southest, at Sand-

[1] The name may be read either Limber or Lunber.

wiche at Davyes gate south and in the Doownys goth half tide under Rothir and yif ye Ride in the Doowns and will go into Sandwiche haven Rere it by turnyng wynde at an est south of the moone, and yif it be a flowyng wynd ye may abide the lenger, and yif ye be bounde to Caleis haven and Ride in the Doowns, and the wynde be west south west, ye must Rere at a North north est moone and gete you into your merkis. the steple into the fan, than go your cours Est south-est ovir and aftir your wynde and your tide serve your cours. And loke ye seeke Caleis haven at a south southest mone or els at a South and by est. And yif ye turne in the Downes come not nere Godwyn than ix. fadome ne not nere the brakis than v. fadome. Fro Seint Margret steyers and ye will go with Dengenes, your best way is south south west and seke you xviij. fadome depe be twene seint Margaret steyers and Dengenesse goeth half tide, and fro Dengenes to Hildirnes your cours is Est and West Dengenes. and the watir of Sowm lyeth est south est and west north west, Dengenes and depe southest and by est Northwest and by west, Dengenes and ye have xx$^{ti.}$ fadome depe. Westsouthwest and est north est that is your cours along the see, and at Dengenes is half and half quarter tide and south unto Hastyngis half tide as by cheffe quarter tide Be chif and Depe south est and northwest. Bechif and the Seyn hede south and north, the Ile of Arundele and Strotarde south southest and north northwest the Seyne hede and Wolneshorde[1] south est and northwest Berfletnes and Wolneshorde south southest and north north west The Chapell of Hoggis and the Nedles south and north. the Hagge be est Rokesnes and Wolneshorde south and by west north and by est Wolneshorde and Garnesey southe southwest and quarter tide at Wolneshorde. Fro Wolnes-horde unto the Ligge of Seint Elenes is half tide undir Rothir. And from Seint Elenes to Chakkeshorde is half tide

[1] The name may be Wolveshorde, as there is no difference in fifteenth century writing between the letters *n* and *u*, and the latter continually stands for *v*.

and a south moone makith high watir within Wiet the nedlis and the forne lieth south west, and by west north est and by est the nedlis and Cornelande est and west. At the nedlis it flowith south est and by south fro the nedles to Portlonde the cours is west south west and est north est at the Polketh in haven it flowith northwest and southest, and in the fairway south southest and north northwest. At Waymouth within havyn Est and West at the Bill at Portlonde south south est and north northwest, the Seyne hede at Portlonde and Garnesey south and north, Seyne hede and the hay wode be west Dertmouth est and by south west and be north, Abottysbury and the forne lieth northest and south west, Portlande and bery land is est and by north west and by south. bery laund and the Stert west south west and Est northest, betwene Portlande and the Stert ever (?)[1] havyn is tide est and west betwene Bery londe and the Londis ende of Englonde there is half tide. In the fairway betwene the Start and Lisart the cours is est and west. And beware of the hidre stonys. All the havens be full at a west south west moone betwene the Start and Lisart, the Londis ende and Lisard lieth est southest and west northe west. At the Londis ende lieth Raynoldis stone. A litill birth of but xij. fadom shall lede you all be owten hym and south south west of the Landes ende lieth the gulf. the langshippis and the landende lien north northwest and south southest. And it flowith west southwest and half tide undre Rothir by londe, but none the long shippis and seint mary Sande of Cille lith west south west and est northest Seint Mary of Cille and Uschante lien northwest and by north south est and by south Cille and the seyne lien south southest and north northwest, the seyne and Huschaunt lieth south and north Huschaunt and the pople hope lien north and by west south and be est Huschaunt and Lisard north and south, Lisarde and seint

[1] *Ever* or *euer*. The reading may be "Stertener" intended to be read as one word, though written like two.

Mary sande of Cille est and west but beware the gulf. Saint mary sande and the forne northwest and southest the forne and the poplehope north northwest and south southest the forne and lisard north and by west south and by est. the forne and the grey be est. Falmouth north and south. the forne and the Ram hede south southwest and north north est, huschaunt and the Ram hede northest and by north southwest and by south, be forne and berylonde north est and by north south west and be south, the start and baspalis north and south, baspalis and the Ramhede north and by estė South and by west, Garnesey and the hey wode be west Dertmouth west north west and est southest. In spayne and bretayne this is the cours and the tide. fro Seint Maluys unto baspalis the cours is est northest and west south west, and open of baspalis lieth the langas it flowith Est and west on the cooste. the Langas and the estbrigge lye south and by west northest and by Est till ye come into your fairway yif ye be bounde Estwarde ye shall go north est, and yif ye be bounde westwarde ye shall go west southwest till ye com ayenst the forne. At the forne goth half tide betwene Huschaunt and thee forne the Cours of the Chanell of Seint Mathyus and ye go withoute the bradreth ye must go for to go clene of all daungers your cours is south and by est north and by west, but wynde makith cours. And at Seynt Matheus it flowith Est north est and south southwest. At the forlande of fontenes it flowith southwest and northest, but a man that ridith in the way of odierene at an ankre, he may begyn to rere at an est southest moone for to turne And the wynde be at the north est or hou so evir it be fro the forelonde of fontenes to the straitis of Marrok. A south west mone makith hiest watir by the see coste, and in the updraughtis it dooth not so the forelonde of fontenes and penmarke lien north west and southest And Penmark and the saine north west and be west south est and be est the saine and by Huschaunt north and south Penmarke And be like west north west and est south est, beware of Vas glenaunt the streeme settith southwest and north est go fro the saine southest and by est, and ye bee in lx. fadome depe and x and ye shall fall with eleron, than go your cours with the pelehede south est and by south and ye be in xij. fadome depe, And than shall lede you w^toute the poullis. Fro the Pelis ye must go est north est till ye be above the piper, than go est and by north for cause of the horshoo. And than ye may go from opyn on the blake shore est southest till ye come as high in geronde as talamont, for the groundes on the southir side lyen ferr oute, and arne shore too, for ye may come no nere them than vij. fadome. And when ye come anens talamont ye shall go with Castillion south south est And beware of the mydill grounde use and be lile lien south est and northwest be like and the pekelerre lye west northwest and est south est the tutport and the pelis lyen west northwest and est southest the pelis and the borugh of baion south and north. go fro the pelis of Amians west southwest, And go clene of all the coste of Spayne and ye shall come by Siete of Cap' finestre all high up use and macheschaco southest and by south northwest and be north belile and seint Tony south and north. belille and seint Andrews north and by est south And by west belille and Ortingere southwest and by south north est and by north Belille and the Cap' fenister southwest and by west north est and by est the saine and the bokowe of Vaion south est and northwest Maschechaco and Sayne southest and by south northwest and be north the sayne and seint Tony south south est and north northwest, Seint Andrews And the Seyne north and by west south and by Est Seint Sebastians and the saine south and north. Ortinger and Huschaunt sooth south west and north northest the forelonde of fontenes and the cap' Fenistre northest and south west be ware of the saine fro the bokowe of bayon to the cap fenistre the fair way is est and west, the cap fenistre and the berlinge sooth and north the birlingis and the Rokkes seynter south south est and north northwest cape seint Vincent and cape seint marie est and by south west and by north cape seint Marie and Caleis maly southest and by est north west and by west Calus and the River of civell south est and by south north west and by north, Cape seint marie and the straitis south est and northwest the straites est north est and west southwest, Cape fenister and mews nesse north and by west south and by est Cape fenister and clere in Irlonde north and South cape fenistre and cille north north est and south south west clere and the bokowe of baion southest and northwest clere and seint Tony in Spayne north north west and southe southest clere and Ortingere north and by west south and by est clere and the saine est southest and west northwest clere and cille south est and northwest, cille and the holde hede of Hinderfforde south yest and by est north west and by west A newe cours and tide betwene Englonde and Irlonde the Londis end and the holde hede of Hinderforde west north west and est southest shipmanhede of cille and the seven stonys southest and northest, the long shippis and the vij. stonys est and west the Londis ende and the Yokelis north west and southest the Londis ende and the toure of Watirforde north northwest and south south est, the toure of Watirforde and the toure of Velafade north and south, the Londis ende and saltais north and by west south and by est Tuskarde and long shippis north and south freston herde and smal of skidwale north and by est south and by west Ferstonhorde and seint Thomas forlande on the west side of Milforde north north est and soth south west, est and west it flowith within the havyn and half streme vndir Rothir and w^{t}oute it renneth north est and southwest, shipmanhede and mylford north est and by north south west and by south. Shipmanhede and Londay north est and southwest. And be ware of the vij. stonys Frestonhorde and Londay north est and by north south west and by south. Londay and Calday north and south. fro the Londis ende to Londay it flowith west southwest and est northest fro Londay to the Holmys est and by north west and by south be ware of Iron groundis and of your stremes of flode for they sitten north est on the Iron groundis. And on ebbe spare not to goo for the streemys of Briggewatir sit west norwest. And beware of Columsonde it flowith fro Londay to the Holmys est and west and fro the Holmys for to go clene of the Wasshe groundis and of Longbors the cours is north. And ye come on ebbe and sith go est and north est with Portis hede but yif ye have a quarter tide at the flat holme ye may goo est north est or est and by south and go ovir Langborde with Ketils wode with a gode ship, for ye shall have iij. fadome on the sonde or more by that ye come there betwene the holmys and Ketilsworde and Portishede it flowith west, and by north est and by south at Kyngrode it flowith est and west. And set on no lesse watir above the holmys than xij. fadome at the leest, Seint Thomas forlonde and stalmay lieth north-west and southest. All that see betwene Irlonde and Walis goth half tide under Rothir, londay and the old hede of Hindilforde lye west and by north and est and by south. And yif ye be bounde[1]

go west northwest. And ye shall go clene of Kidwall and small and ye have any ebbe the streme settith north north est and south southwest. And there is half tide undir Rothir for it flowith on londe est and west, fro tuscarde to the olde hede of Hindilforde to Clere in Irlonde the cours is west and by south est and by north, Clere and mews nesse and thursay north west and southest, thursay and the lewe north northwest and south southest, the sowde of blaskay lye north and south, blaskay and the Ackiles north and south. Blaskay and the stakis of Connothe north north-est and South southwest, but thou must go north and by est for a Rokke the stakis of Romney. And the Londes end of

[1] Here half a line is left blank in the original MS.

Irlonde north northest and south southwest. And so thou must go to the Ilonde of Torre the stakis of Conney and southwest and northest. And fro the stakes of Conney to the legge of Rabyn the cours is west southwest and est north est, the sonde and the forelonde be est Loswill lieth west southwest and est north est, but be ware of the Rokke in the Bay of Loswill. Fro the forlonde of Loswill to Donsmares hede the cours is west north west and est southest, the sounde of Ranseynes the same cours with Benoster fro Tuscarde to Donsmere hede it flowith by the see cost west south west and est north est, But in the updraughtis it dooth not soo, fro tuscarde to the redebank it is half tide undir Rothir. Fro saltais to tuscarde the cours is est and west, fro the tuscarde to the hede of the skarres for to go clene of all the gronde betwene tuscarde and Dalcay the cours is north est and south southwest, fro the Skarris unto Arglas the cours is north and south, Fro Arglas ye shall go with Capman eylond south southest and north northwest but and ye be bounde to Capman ylonde ye shall go north and by west, for cause of ij Rokkes that lien in the wey. And yif ye be bounde south warde ye shall go south est and by south. Fro Capman ylonde to the forlonde of Welnerferth ye shall goo north northest and south southwest, fro the forlonde of Wolnderfrith to benestore south southest and north northwest it flowith on the coste betwene tuscarde and beneford, south southest and north northwest, betwene Capman Ilonde and Donblak. And by south Arglas there goth quarter tide undir Rothir Capman ylonde and the Ile of Man the southende lieth south and by est northwest. And by west the Ilonde of Man and Arglas Est north est and west south west, the Ile of Man and Lambey Ilonde north est and south west, the Holbe and the Holy hede est and west Lambay and tha Ramsair north and south, the chirch of Wiklowe and the Ransires south southest. but a man that ridith in the Rode of Wiklowe must go oute of the chirch of Wiklowe south est and northwest, Tuscarde and the Ransere est and west, the toure of Watirford and gresholme west and be north est and be south. All that see goth half tide betwene the smale and Skidwhalles and the bersays. And it flowith est and west on the mayne londe and at at[1] the Ramseir north and south the stremys renne in the sonde and be owten the Bisshoppis and his clerkis north northwest and south south est, sculke holme and the sonde of Ramseirs north and south And beware the Rok men callith Sampson for he lieth at the south point at seint Davy side. And kepe more nere the Ilonde than the mayne londe till ye be passid the point and thorowe the sande, than go north till ye come at a nothir Rok. And for cause of that Rok ye must go north and by west or els north and by est for north is even with the Rok. And the name of the Rok is called the Kep', and he lieth undir the watir but it brekith upon hym And the breche shewith, And than your cours is north northest for to go with barseis stremys, and seint Davies londe northest and southwest. And so go your cours north northest and south south west till ye come to Ire north west upon Scotlande the Holy Hede and the Ile of man north and by est south and be west. And yif ye go to Chestir ye shall go fro the scarris till ye come anens the Castell of Rotlonde your cours is west southwest and est northest. And take your saught on the mayne londe of Wales Rotlonde and the Redebank in Chestre watir north and South.[2]

Opyn oo grounde there is wose and sonde togidir and it is bein xij. or xiiij. fadome or xvj. fadome depe. Upon opertus Mamoschaunt there is stynkyng wose and xij fadome depe. opon opertus antiage there is blake sande opon o the taile of ars is xxiiij$^{ti.}$ or xxvj. fadome depe there is grete grey sonde and smale blake stonys and grete whit shellis among upon of use there is l. or lx. fadome depe wosy sonde. Open

[1] *Sic.* "at" repeated.

[2] Here a small space is left blank in the MS.

one Liere there is stremy grounde and white shellis. upon o belille there is in lx. fadome or lxx. smale diale sonde Opyn of Penmarke there is in l. fadome blak wose Opyn the same in lx. fadome there is sandy wose and blak fischey stonys among Opyn of Huschaunt in l. or lx. fadome there is redd sande and blak stonys and white shellis among betwene Cille and Huschant there is grete stremy grounde with white shellis among withoute Cille west south west of hym the grounde is Rede sonde and white shellis amonge, between Cille and Lesarde the grounde is white sonde and white shellis shellis[1] Among Opyn Lesarde is grete stone as it were benys and it is raggid stoon Opyn of Dudman in xl fadome there is rede sande and whit shellis and small blak stonys amonge Opon oporte londe there is feir white sande and xxiiij$^{ti.}$ fadome with Rede shellis therein, And in xiiij. or xvj. fadome there is rokky grounde and in sumplace there is feir cley grounde Opon a Wiet there is fere hard platmer grounde. And the faire way in xxx$^{ti.}$ fadome there is white chalky grounde Opyn o bechefe there is sande and gravell to gidir in xx$^{ti.}$ fadome depe. Here be the groundis of Inglonde bretayne and Cille. And ye come oute of Spayne. And ye bee at capfenister go your cours north northest. And ye gesse you ij. parties ovir the see and be bounde into sebarne ye must north and by est till ye come into Sowdyng, And yif ye have an C. fadome depe or els $\overset{xx}{\text{iiij}}$.x. than ye shall go north in till the sonde ayen in lxxij. fadome in feir grey sonde And that is the Rigge that lieth betwene clere and Cille than go north till ye come into sowdyng of woyse. and than go your cours est north est or els est and by north and ye shall not faile much of Stepilhorde he risith all rounde as it were a Coppid hille. And yif ye be three parties ovir the see and ye be bounde into the narowe see and ye go north northest and by north till ye come into sowdyng of an hun-

[1] So in MS., repeated.

drith fadom depe than go your cours north est till ye come into iiij^{xx} fadome depe. And yif it be stremy grounde it is betwene Huschaunt and cille in the entry in the Chanell of Flaundres. And so go your cours till ye have lx. fadome depe. than go est northest along the see, etc.

INTRODUCTORY REMARKS TO GLOSSARY.

" Gentyll maryners on a bonne vyage,
Hoyce vp the sayle, and let God stere
In ye bonauenture making your passage.
It is ful see the wether fayre and clere,
The nepe tydes shall you nothing dere,
A see bord mates S. George to borow,
Mary and John, ye shal not nede to fere,
But with this boke to go safe thorow."

(The Rutter of the Sea—Prologue.)

The curious treatise printed in the foregoing pages came into the possession of the Hakluyt Society in 1880, through Mr. Gairdner, of the Public Record Office, who had it transcribed for the Camden Society. Finding its interest, however, to be purely geographical, and therefore more suitable for a Society like ours, he transferred it, together with his prefatory remarks, to my predecessor, Mr. Clements Markham. The printed sheets have been lying by ever since, waiting an opportunity of incorporating them with some other kindred work. Such an opportunity has at last been afforded by the issue of the present volume. But in order to make these old sailing directions intelligible to our readers it was obvious that some kind of a commentary was necessary. This I have attempted in the accompanying glossary, and have added a map on which the names of places are marked in their old and modern form.

In identifying the names of places, the following works have been consulted: *The Lighting Colomne, or Sea Mirrour*, by Peter Goos, printed at Amsterdam in 1658; *The English Coasting Pilot, or Sea Mirrour*, by Casparus Lootsman (*i.e.*, Caspar the Pilot), also published at Amsterdam in 1693; Seller's *Coasting and English Pilots* (1670-1680); Grenville Collins' *Coasting Pilot*, 1693; *A Description and Platte of the Seacoast*, author unknown, printed in 1653; Ortelius' *Atlas*, 1570; Saxton's *Atlas*, 1579; Imray's *Sailing Directions*, Norie's *Sailing Directions*, revised by

Hobbs; Burat's *Côtes de France;* Camden's *Britannia;* and the *English and French Admiralty Charts.*

Many of the names appeared, at first sight, hopelessly difficult, and it was only after patient investigation and research that their meaning became clear—for who would suspect that "Leyrnes" referred to the well-known town of Wainfleet, or that "the Shelde" was no other than the now fashionable sea-bathing place of Cromer; that "Whitvies", "the Spone", and "Wolveshorde" were respectively Whitby, the Spurnhead, and Dunnose Point? Passing to the other side of the English Channel, or the Channel of Flanders, as it was then called, we find such names as "the Hagge" for Cape La Hague, "Hoggis" for Cape La Hougue, "Berfletnes" for Cape Barfleur, and many other curiosities. Turning to obsolete terms, "Undir Rothir" occurs several times, and always with reference to tides. We have, "At the Shelde (*i.e.*, Cromer), it floweth on the londe westnorthwest and half streme (stream) *undir Rothir* by the londe till ye come to Winterbornes (*i.e.*, Winterton ness)"; and again, "from Wyntirburnes till ye coome to Cukle rode (Kirkley road) it flowith on the londe northwest and quarter tide and half quarter *undir Rothir.*" So, again, in the Downs we are told it goeth "half tide *undir Rothir*".

This expression "under Rothir" presented considerable difficulty. The Dictionaries threw no light upon it, but rather led me off the scent by giving "Rothir", an old form of "rudder"; and many were my attempts to account for the tide running differently under the rudder from what it did under any other part of the ship.

At the British Museum Library, however, I came upon a little book entitled *The Rutter of the see with the hauens rodes, soundynges, kennynges, wyndes, flodes and ebbes, daungers and coastes of dyuers regions, with the lawes of the yle Auleron and the iudgementes of the see, with a Rutter of the Northe added to the same.* The first part of this work is a translation by Robert Copland, a pupil of the famous Caxton, of a French *routier* (Angl. rutter). The last part, compiled by Richard Prowde in 1541, is a reproduction of our "Sailing Directions", breaking off at Dartmouth. No clue is given as to the true authorship of the treatise by the compiler, who merely associates his own name with it. I am inclined, however, to attribute its origin to Clement Paston of Oxnead, Norfolk, a great navigator in the time of Henry VIII. He was fourth son of Sir William Paston, and distinguished himself in the wars of that period.

On comparing this printed version with our transcript I find the words "Undir Rothir" rendered "under other"; and in William Bourne's *Regiment of the Sea,* a sixteenth century treatise on navigation, directions as to tides are also followed by the words "under the other". Thus we are told (leaf 14, back), that "from Fairely to Be(a)chy (Head) it runneth quarter tide *under other*"; and on leaf 15, "It floweth all alongst the coast of Flanders from the Wildings to Calys (Calais), a south and by East moone; and so runneth halfe tide *under the other*."

The meaning of "under Rothir" now becomes clear, for the late Sir George Airy, in his treatise on "Tides and Waves" in the *Encyclopædia Metropolitana,* says that the tides in the English Channel claim notice as having been the subject of careful examination by many persons, English and French. It appears that in the upper part of the Channel the water flows up the Channel nearly three hours after high water and runs down nearly three hours after low water. This continuance of the current after high water, if it last three hours, is called by sailors "tide and half tide"; if it last one hour and a half, it is called "tide and quarter tide". It is obvious that the tidal currents are then flowing in opposite directions, one under the other, and thus we have a satisfactory explanation of the term "under Rothir", without following up the intricate subject of tides any further. (Cf. *Manual of Tides and Tidal Currents,* by the Rev. S. Haughton.)

The identification of Cromer with "the Shelde" of our MS. was another difficulty, for although the names occur together in two old "Sailing Directions" translated from the Dutch, in one marked on a map as "Dager and *Schild*" on the coast, a little to the north-west of Cromer, in the other it occurs in the text as follows: "From the poynt of Cromer or *Schield* to the Tessel (*i.e.,* Texel) the course is East"—yet in none of the English charts, maps, or coasting pilots does the name Shelde or Shild appear near or with reference to Cromer, nor from inquiries made on the spot could I learn of any such name having been connected with the place. Possibly "Dager and Shild" may have something to do with the "Dogger bank", and Dutch navigators in those times may have shaped their course from a point a little above Cromer in order to pass safely between that dangerous shoal and Well Bank to the south of it in crossing to Holland. However this may be, the fact remains proved that "the Shelde" of our MS. is identical with Cromer, a place of some maritime importance up to the middle of the 16th century. (Cf. *History of Norfolk*, by W. B. Rye, p. 250.)

With regard to another identification, "Ile of Arundele", I have endeavoured to show how Arundel might have been in early times an island. On referring, however, to Richard Prowde's version, I find that he has "Hiland (High land) of Arundel" in the same passage. This of course throws quite a different light on the words. It may easily be imagined that, through ignorance or carelessness of the transcriber, "Hiland" may have become changed into "Iland", and this again into "Ile". I have, however, allowed my glossary note to stand, so that the reader may decide the point for himself.

In conclusion, I beg to acknowledge, with thanks, kind advice and suggestions received from Dr. Richard Garnett of the British Museum, and from Admiral Brine.

E. Delmar Morgan.

GLOSSARY.

Abbotysbury—Abbotsbury, on the coast of Dorsetshire.

Ackiles, The—Achil Head, on Achill Island, off the coast of Mayo.

Anens—Against, opposite.

Antiage, pertus—Pertuis d'Antioche, the passage between Ile de Ré and Ile d'Oleron, leading to Rochelle. The passage takes its name from some rocky banks called the "Antioches".

Arglas—Ardglass, east coast of Ireland, a few miles above St. John's Point.

Ars, Taile of—Pointe d'Arseaux, now called St. Martin's Bank, extending eastward from Ile de Ré to the middle of the channel leading to Rochelle. Ars steeple was one of the marks for the navigation of these waters.

Arundele, Ile of—Old charts represent Arundel on a peninsula, with its promontory stretching far out seaward, and the wide estuaries of two rivers, the Arun and Adur, on either side. This probably explains the term "Isle of Arundel". "We must bear in mind", says a writer in the *Sussex Archæological Collections* (vol. xi, 93), "that the whole of the levels of the river Arun were covered by water every tide, and not confined to a narrow channel as now, and that to facilitate a passage through this valley without interruption at all times a causeway was thrown up its whole width. . . ." Arundel itself, the *ad Decimum Lapidem* of the Romans, was originally a British town, with the river on one side, a marshy and wooded ravine on the other, and a *fosse* and *vallum* traversing the neck of land between the two. Arundel, now some distance inland, was a seaport, and is spoken of as "eminent for building ships", the forests in the vicinity supplying the material. (*But see* Introductory Remarks.)

Ayen—Again.

Baion and **Vaion**—Bayonne, at the confluence of the rivers Adour and Nieve, in lat. 43° 29′ 15″ N., and long. 1° 28′ 17″ W. from Greenwich. It contains 70,000 inhabitants, and is the chief town in the department of the Lower Pyrenees.

Baspalis (Ile de Bas)—An island off the north coast of Brittany ; the tide here rises and falls nearly thirty feet, covering half the island at flood. Hence its name of "Low Island".

Be like and **Be lile**—The island of Belle Ile, between 9 and 10 miles long and 3 or 4 miles broad. This island is lofty and steep, spacious and fertile, and its deeply indented coast affords shelter and anchorage to navigators. Its name therefore is appropriate.

Benoster—Probably Benmore, or Fair Head, north-east coast of Ireland.

Benys, *i.e.*, beans—In old sailing directions we find "Great rough stones as big as beans".

Be owten—Without, in the sense of outside.

Berfletnes—Cape Barfleur.

Berlinge and **Birlingis**—The Burlings rocks, off the coast of Portugal, in lat. 39° 25′ N., long. 9° 28′ W.

Bersays and **Barseis**—Barsey, or Bardsey Island, off the coast of Carnarvon, 70 miles N.E. ¾ E. from the Small's lighthouse, and 20 leagues E. ½ N. from the Tuskar rock. A channel 1¾ mile wide separates Bardsey Island from the mainland.

Berwik—Berwick, a fortified town on the Tweed, one of the principal sea-ports in Scotland in the 12th century. In 1482 it came finally into the possession of England.

Bery land—Berry Head, on the south side of the entrance to Torbay.

Birth—Berth, "a litill birth" would, in sailors' parlance, mean a wide distance.

Bisshoppis and his Clerkis—The Bishop and his clerks, a number of dangerous rocks lying N.W. of Ramsey Island, off the coast of Wales.

Blake shore, The—Terre Negre, on the south shore of the entrance of the Gironde ; a fixed light now stands here.

Blaskay, The sowde of—Blasket Sound, west coast of Ireland.

Bokowe—From the Italian *bocca*, mouth or estuary of a river.

Borugh—*i.e.*, borough or town.

Bradreth, The—Brest Sound.

Brakis, The—The Brake sand, 4½ miles long, between the North Foreland and the Downs. This shoal is marked by three buoys, north, middle, and south Brake, known collectively as "the Brakes".

Breche—Breach, in the sense of breakers.

Bretayne—Brittany.

Briggewatir—Bridgewater, in the Bristol Channel.

Bycheffe and **Bechif**—Beachy Head, the remarkable headland with its high chalk cliff, 9½ leagues W. ¼ N. from Dungeness.

By in—Within.

Calday—Caldy Island, north of the entrance to the Bristol Channel.

Caleis Maly and **Calus**—Cadiz.

Cap' finistre and **fenister**—Cape Finisterre.

Capman eylond—Copeland or Copland Island, 2 miles N.E. ½ N. from Donaghadee, east coast of Ireland.

Castillion—On the south or Médoc shore of the Garonne, the modern Castelnau.

Chakkeshorde—Probably Chichester (also called in old sailing directions Chaikeshord) ; the termination "horde" is merely the German "ord", modern German "ort", a place.

Chestir—Chester.

Cille—The Scilly Isles.

Civell, River of—River of Seville, or Guadalquivir.

Clene—Clean, *i.e.*, clear.

Clere—Cape Clear, the southernmost point of Ireland.

Columsonde—The Culver sand, a dangerous and extensive flat to the northward of Bridgewater ; a narrow ridge of this sand dries for the extent of 3 miles, with long spits at each end.

Connothe and **Conney, The Stakis of**—The Stags of Connaught, some rocks off Broadhaven Bay, county Mayo, west coast of Ireland.

Coppid—Topped, in the sense of overhanging masses of rock, from " Cope", whence our word " coping", *e.g.*, coping brick.

Cornelande—Cornwall, the horn-shaped land ; the ancient name for this county being *Kernou* or *Kerniw*, the Horn, from its projecting promontories.

Coste—Coast.

Cukle rode—Cockle Gat, the passage forming the entrance into Yarmouth Roads, and now called Nelson's Gat. A light vessel is moored here.

Dalcay—Dalkey Island, south of Kingstown, east coast of Ireland.

Dengenes—Dungeness, also written in old sailing directions " Dongie Nesse."

Depe—Dieppe.

Dertmouth—Dartmouth.

Dial sonde—Fine sand, suitable for hour-glasses.

Doownys, The—The Downs.

Donblak—Dundalk Bay, county Louth, east coast of Ireland.

Donsmares hede and **Donsmere hede**—Dunmore Head, north coast of Ireland.

Dudman—Deadman Head, east of Veryon Bay, Cornwall.

Eleron, The—Ile d'Oleron, off the coast of the Charente, opposite the entrance to Rochefort. Oleron was also known for its laws, a body of rules for the guidance of maritime cases. These were translated into English and published about 1540. (*See* Introductory Remarks.)

Estermare cours(e)—*i.e.*, the course for sailing to the North Sea and coast of Holland.

Ever and **euer**—For " every".

Fan—Probably for vane or weathercock.

Feir and **Fere**—For " fair".

Fenyn Ilonde—Ferne or Farne Island, the largest of a group of rocky islets E. by S. 2 miles from Bambrough Castle.

Flaunders, Chanell of—The English Channel.

Flode, On—Floodtide.

Fontenes, Forlande of—Point, or Bec, du Raz, on the coast of Brittany. On its highest part stands a lighthouse, which may be seen in fine weather at a distance of six leagues. The Abbey of Fontenay is mentioned in Exchequer Rolls of the 14th century.

Forne, The—The Four, or Oven, a remarkable black rock never covered, about a mile from the north-west point of Brittany, and ten miles from Ushant lighthouse. The Passage du Four, between Ushant and the mainland, takes its name from the rock.

Garnesey—The Island of Guernsey.

Geronde—The Gironde, or Garonne, the river of Bordeaux. Many towns and villages are situated on its banks, such as Pauillac, Blaye. But its navigation is so dangerous that vessels are advised not to enter it by night and in thick weather.

Gesse, Till the (ye), *i.e.*, till ye guess.

Glas or two, A—Evidently referring to the hour-glass, an important accessory in navigation up to a recent period. Clocks and watches were in use in the 15th century, or earlier on shore, but it is uncertain when they were first used at sea.

Godewyn—Goodwin.

Gold stonys—Gold Stone, a dangerous rock, rather more than a mile E.S.E. from Holy Island Castle. It is very small, and visible at low water. In old sailing directions "the Plough", another sunken rock near it, was generally included in the term "Gold stones".

Grene bank—Probably the Isle of Grain, at the mouth of the Medway.

Grey, The—Probably St. Michael's Mount. The Cornish name for this isolated rock in Mount's Bay, was *Caraclowse* or *Careg Cowse*, the Gray or Hoary Rock; and Camden says the inhabitants called it so.

Gresholme—Gresholm or Grassholm Island, south of St. David's Head, South Wales, and usually the first land seen on coming towards Milford Haven from the westward.

Gulf, The—A rock S.S.W. from the Land's End, and 5 leagues E. from Scilly, marked in modern charts as "the Wolf".

Hagge, The—Cape La Hague, the headland of Normandy, opposite the Island of Alderney. It forms the north-west extremity of the peninsula of Cotentin, in the department of La Manche.

Hastyngis—Hastings.

Hay wode—Hyant wood, one of the marks for sailing into Stoke's Bay. Havant, on the coast of Hampshire, possibly takes its name from it.

Hedelonde, The—Flamborough Head, the well-known promontory on the Yorkshire coast.

Hidre stonys—Hidden stones, in the sense of sunk rocks; possibly our word "eddy" is derived from this old form of "hidden".

Hildirnes—Probably Cape Grisnez, on the French coast, formerly known as Whiteness and Blackness.

Hinderfforde, Holde hede of—The old Head of Kinsale, south coast of Ireland.

Hoggis, Chapell of—Cape La Hogue or La Hougue, on the coast of Normandy, with Capelle Road a little to the south of it. Here, in 1692, the French fleet was defeated and almost destroyed by the combined English and Dutch fleets.

Holbe, The—Probably Bantry Bay.

Holdernes—Holderness, the low-lying south-east corner of the East Riding of Yorkshire, terminated at the extreme point by Spurn Head.

Holmes, The—The Holms, a large sandy flat at the entrance into Yarmouth Roads.

Holmys, The—The Holms, two small islands lying some distance apart, but nearly in the middle of the Bristol Channel. The southernmost of the two is called Steep Holm; the other, about 2 miles from it N. by E. ½ E., is Flat Holm.

Holmys hede—The head of Holm Sand, off Lowestoft.

Holy hede—Holyhead.

Horseshoo, The—The Horse-bank and Horse-shoe Hole, an anchoring ground between the Nore and North Foreland.

Horshoo, The—Probably a bank of sand at the mouth of the Gironde. The two banks which front the river are now called *La Mauvaise* and *La Cuivre*.

Houndeclif fote—Huntley Foot, marked Huntcliff Foot on seventeenth century charts.

Hushaunt. (*See* **Uschante.**)

Ilonde, The—Holy Island, or Lindisfarne, about 1½ mile from the mainland. The course and distance from Bambrough Castle to the south point of Holy Island are N. ¾ W., 4¼ miles.

Ire—Point of Air, S.E. by E. ¾ E., distant 19 miles from Great Orme's Head, at the entrance to the river Dee; or, more probably, Point of Ayr, the northernmost point of the Isle of Man.

Iron groundis—Probably referring to the iron-bound rocky coast south of the Bristol Channel, extending for 24 miles eastward from Ilfracombe.

It flows tide and half-tide—According to the *Seaman's Grammar*, this means that it will be high water sooner by three hours at the shore than in the offing. (*See* Introductory Remarks.)

Kep', The—Probably a rock, "the Keep"; according to our text, its position would be due north from St. David's Head.

Kidwall. (*See* **Skidwale.**)

Knak, in the Kentisshe Sea, The—The Kentish Knock, a dangerous and extensive shoal, about 19 miles N.E. ½ E. from the North Foreland lighthouse. Its length is about 7 miles, and its broadest part 2 miles.

Kirkleholmys—Kirkley Holms, off Lowestoff.

Kyngrode—King Road, between Portishead and Bristol.

Lambey Ilonde—Lambay Island, county Meath, off the east coast of Ireland, 7 miles from Howth Head. The name is probably from lamb, the animal.

Langas, The—Probably the Tour de la Lande, a leading mark for entering the Morlaix river from Ile de Bas.

Lang shippis and **Long shippis**—The Long ships, a group of rocks lying about 3 miles N.N.W. ½ W. from the south-east point of Land's End; a lighthouse now stands on the highest of them.

Lewe, The—Loop Head, west coast of Ireland, north of the Shannon.

Leyrnes and **Leirnes**—Winfleet or Wainfleet, on the coast of Lincolnshire. In seventeenth century sailing directions this place is referred to as Legerness and Lagerness.

Liere—Leyre, river and bay in Côte de Landes.

Ligge—A low-lying spit of land.

Limber and **Urry**—The Leman and Ower, two dangerous shoals lying off the coast between Foulness and Flamborough Head. These shoals are buoyed, and a light vessel is moored between them.

Lisart and **Lisard**—Lizard Point, the southmost part of England, a bold-looking land, seen in clear weather 20 miles off.

Londay—Lundy Island, off the entrance to the Bristol Channel.

Londes end of Irlande—Ireland's north point, near Malin Head.

Londis ende—Land's End, the westmost part of England.

Longbors and **Langborde**—Probably a shoal in the Bristol Channel.

Long Sande—A shoal 15½ miles long, off the mouth of the Thames.

Loswill—Lough Swilly, north coast of Ireland.

Macheschaco—Cape Machichaco, on the north coast of Spain, now marked by a lighthouse.

Mamoschaunt, Pertus—Pertuis Maumusson, the south passage between Ile d'Oleron and the Charente; a dangerous channel, little known except to the natives.

Marrok, Straitis of—The Straits of Gibraltar.

Maylaunde—Mainland.

Merkis—Marks, *i.e.*, leading marks used in navigation.

Mews nesse—Mizan Head, south-west coast of Ireland.

Milforde and **Mylford**—Milford haven.

Mydill—Middle.

Naisse and **Nasse**—The Naze or headland of Essex, south of Harwich.

Nedles—The Needles rocks and point at the west end of the Isle of Wight.

Ne nere—Nor nearer. Cf. the use of *ne* for *nor* in "The Childe of Bristow", an early poem, published in the *Camden Miscellany*, vol. iv.

Odierne, The way of—Audierne or Hodierne Bay, is a slight indentation of the coast between Fontenay, Raz de Sein, and Penmark Point. The harbour of Andierne is tidal, and can only be entered at high water.

Open of, opyn on, and **opyn ou**—On, upon.

Opertus—The letter *o* stands for the preposition "on", the remainder of the word, sometimes spelt "porthus", is an Anglicised form of the French *pertuis*, a narrow passage between an island and the mainland, as in Pertuis d'Antioche, Pertuis Maumusson. The word is derived from the Latin *pertusus*, participle of *pertundere*, to pierce, from *per* and *tundere*, and is distinct from the Lat. *portus*, whence our "port" and the Celtic "porth".

Oporte lande—*O*, the first letter, is the preposition "on", the remainder being Portland in Dorset.

Ortingere—Cape Ortegal, in lat. 43° 46′ 30″ and long. 7° 48′ 15″ W. from Greenwich. A watch-tower is built on the summit of the cape, affording a good mark to vessels making the land.

Orwell haven—The harbour of Harwich, formed by the junction of the Stour and the Orwell.

Orwell waynys—Orwell wains or wands, at the entrance to Harwich.

Pekelerre—Picquelier Island, off the promontory of Armentier, in Poictou.

Pele hede, The—The Pole head, at the entrance of the Gironde, or river of Bordeaux. A lighthouse built on a rock called the tower of Cordouan, stands nearly midway of the entrance, and has long been esteemed the most elegant structure of the kind in Europe.

Pelis and **Pelis of Amians**—Ile du Pilier, a small island off Point de l'Herbaudière, the northwesternmost point of Noirmoutier Island.

Penmarke—Pointe de Penmarch, or Penmark Point, with two groups of dangerous rocks lying off it, known respectively as "Wester Penmarcks" and "Easter Penmarcks", off the coast of Finisterre.

Piper, The—A sandbank in the mouth of the Garonne.

Platmer—Flat, from the French "plat".

Polketh—Polkerris Bay.

Portishede—Portishead, near the mouth of the Avon.

Portlonde, Bill at—Portland Bill, a rocky peninsula projecting from the shore of Dorsetshire, 17 miles west-south-west of St. Alban's Head, and in appearance resembling the beak of a bird, whence its name.

Pople hope—Probably Hope Nose, on the north side of Torbay. *Popple,* in the Hampshire dialect, is a pebble.

Poullis, The—Probably the rocky islands which stud the west coast of France, between Poolquain and the mouth of the Loire.

Rabyn, The legge of—Rathlin Island, north-east coast of Ireland.

Ram hede—Ramehead, the west point of Plymouth Sound.

Ramsair, The, and **Ransires**—Ramsey Island, off St. David's Head.

Ranseynes, The sounde of—Ramsey Sound, between the island of this name and St. David's Head.

Raynoldis stone—Rundle or Runnell stone, a small rock between Mount's Bay and the Land's End, a most dangerous obstacle to navigation. This rock, about 4 yards long and 2 broad, is dry at low water, and covered before half-flood. In a curious account, published in 1590, of the voyage of one Richard Ferris, a Queen's messenger, in a wherry-boat from London to Bristol, the author relates how a pirate lay in wait for him near a rock called "Raynalde stones", and how he managed to escape him by passing on the inner side of the said rock, where, he says, "we went through very pleasantly". (See *Illustrations of Early English Popular Literature,* edited by Collier, vol. ii.)

Rede bank—The Red Bank—(1) a shoal in Chester water; (2) a shoal off the south-east coast of Ireland.

Rere it, *i.e.,* raise anchor.

Re sande—Red sand shoal off the Norfolk coast.

Rigge, The—The Ridge, a rocky ledge at the entrance to Harwich.

Rokesnes—Cape Rokeine, the westernmost part of the Island of Guernsey. Rockain Castle stood here. The bay of the same name presents, says Ansted, a bristling array of rocks stretching out seawards more than two miles, and terminating on the south with the Hanois rocks.

Rokkes Seynter, The—Capo da Roca, on the coast of Portugal, in old works called Cape of Rocksemper and Roxent. Mt. Cintra is immediately to the east of it.

Rokkis, The—Probably the Cliff-foot rocks at the entrance to Harwich harbour, or the West rocks, another group between Court and Long Sand.

Romney, The stakis of—The Stags of Aranmore, rocks off the coast of Donegal.

Rothir—An obsolete form of "rudder". "Rother-nails", with shipwrights, are nails with full heads, used to fasten the rudder-irons of ships. (*See* Introductory Remarks.)

Rotlande, Castell of—Rudland or Rhydland Castle, on the Clwyn or Clwyd, falling into Chester water. The old castle is now a mere shell of red sandstone. It was near Rhydlan that the Welsh, under Caradoc, were defeated by the Saxons under Offa, King of Mercia.

Saine, The—Ile de Sein, or Saint, the largest of a long cluster of islands, rocks, and dangers, which lie in a W.N.W. ½ westerly direction from the Bec du Raz, and are known as the Chaussée de Sein; the island is flat in appearance and low; its inhabitants are chiefly fishermen. A lighthouse has lately been erected on the northern point of Ile de Sein.

Saltais—The Saltees, a group of islands and rocks off the south coast of Ireland, some above and others below water at ebb tide. A light-vessel is stationed here.

Sampson—Probably a rock south of St. David's Head.

Sandwiche—One of the Cinque Ports, and a principal harbour in this part of Kent, ranking next to Hastings in precedence. In the earliest extant sea-song descriptive of a pilgrim's voyage we find—

> "For when they have take the see,
> At Sandwyche or at Wynchelsea,
> At Bristow, or where that hit bee,
> Theyr herts begyn to fayle."
>
> (*Early English Ballads*, printed for the Percy Society.)

Saught—Meaning peace, quiet; the expression "take your saught" would therefore mean "take your rest", the perils of the voyage being over.

Scarris, The—The island and rocks called the Skerries lie about 1¾ mile from Carmel Point, Isle of Anglesey.

Sculke holme—Skokham, a rocky island 4 miles north-west from St. Ann's Point.

Sebarne—The river Severn, or Bristol Channel.

Seint Andrews—Santander, the best harbour on the north coast of Spain, eastward of Cape Ortegal.

Seint Davy Side—St. David's Head.

Seint Elenes—St. Helen's, the easternmost point of the Isle of Wight. St. Helen's, though an inconsiderable place, gives it name to a spacious road in which men-of-war lie. Hassell, in his *Tour of the Isle of Wight*, says there is a large farm in the parish called the Priory, it having been a cell to an abbey of Cluniac monks in Normandy.

St. Hosies—Abbey of St. Osyth's, on the coast of Essex, not far from Colchester. According to Camden, the Abbey was so named after the virgin of royal birth, who was stabbed to death here by Danish pirates.

Seint Maluys—St. Malo, in Brittany. The town stands on a small island, which it completely covers, and is joined by a causeway with the mainland. The harbour is one of the best in this part of France.

Seint Margaret Steyers—Old stairs near St. Margaret.

Seint Marie—Cape Santa Maria, on the south coast of Portugal.

Seint Mary Sande of Cille—St. Mary, the largest of the Scilly Isles. The sound, not sand, as in text, is the best and safest passage into St. Mary's Road.

Seint Matthyus and **Seynt Matheus**—Channel of the passage du Four (*q. v.*), between Ile de Sein and Pointe St. Matthieu on the mainland, at the entrance of the Bay of Brest.

Seint Sebastians—Port San Sebastian, easily discovered by its castle of La Mota, situate on the Mount Orgullo, and its old lighthouse. These are distant from each other about a mile, and may be seen at the distance of 8 or 10 leagues. The town of St. Sebastian is the capital of the district of Guipuscoa, in the province of Biscay.

Seint Thomas forlande—St. Thomas's Head, 1⅝ mile from Weston Head, Bristol Channel.

Seint Tony—Santona, the town and port on the coast of Spain. The hill of Santona, on the northern side of the port, is a good landmark.

Seint Vincent—Cape St. Vincent, the south-west point of Portugal.

Seke up, *i.e.*, fetch, a word used nautically in the sense of to reach, or arrive at.

Seven Stonys—The Seven Stones, a cluster of rocks off the Land's End.

Seyn hede, The—The headland forming the entrance to the Seine, near Havre.

Shelde, The—Cromer, the well-known watering-place on the coast of Norfolk (*see* Introductory Remarks). "Sheld", derived from "shell", in the Suffolk dialect meant pied, of two colours, variegated; hence sheld-apple and sheldrake, a beautifully coloured duck. It is possible that the word may have been applied to Cromer on account of the variegated colour of its sands. Cf. Moor's *Suffolk Words and Phrases.*

Shipman hede—Shipman Head, on Bryer, one of the Scilly Isles.

Siete of, By—Within sight.

Skarres and **Skarris**—Skerries harbour, east coast of Ireland, in county Meath.

Skidwhalles—Probably Stidwall Island, in Carnarvon bay, west coast of Wales.

Slade, The—The Sledway, or fairway channel into Harwich from the east.

Smal and **Smale**—Smalls rocks; a cluster of low and very dangerous rocks off St. David's Head.

Sowdyng—Sounding.

Sowm, Watir of—The river Somme.

Spetis—Spits, banks, or sands, generally projecting from the coast. Those here referred to are probably off Shoeburyness, or Sheerness.

Spone, The—Spurn Head, the point of Holderness, at the mouth of the Humber. In 1677, according to Camden, a lighthouse was built here by one Mr. Justinian Angel, of London.

Stalmay—Scalme, now Skomer Island, south of St. Bride's Bay.

Stepilhorde—Probably Steephill near Ventnor, "horde" being merely a termination having the sense of "place", like the German "ort".

Steple—Probably the steeple of St. Peter's Church, Broadstairs, one of the marks for clearing the Goodwin Sands.

Stert, The—Start point (from the Anglo-Saxon *Steort*, a tail or promontory), a rocky headland in the south of Devonshire.

Straitis and **Straites, The**—*See* **Marrok.**

Stremes of flode—Strong tidal currents; the allusion in our text is probably to the well-known *Bore* in the Bristol Channel.

Strotarde—Struysaert, on the coast of Normandy, north of Havre.

Talamont—Talmont bank forms the eastern side of the channel leading up the Garonne.

Temesse—The Thames.

Tenet—Isle of Thanet.

Thursay—Dursey Head on Dursey island, north-west from Mizan Head.

Tilmouth—Tynemouth Haven, at the mouth of the Tyne.

To gidre and **to gidir**—Together. "Togidir", with the same meaning, occurs in Lydgate's poems.

Torre, Ilonde of—Tory Island, off the north-west coast of Ireland.

Turning wynde and **flowing wynd**—A ship is said to be "turning" or beating to windward with a head wind, a "turning wind" would, therefore, be contrary to the course to be sailed; a "flowing wind" would be abeam when a ship could sail with a flowing sheet.

Tuskarde—The Tuskar, a remarkably high rock, 20 feet above water at high tide, lying E.S.E. ½ E. of Carnsore point, and 43 leagues N. by E. ¾ E. from the Longships lighthouse.

Updraughtis—Probably the same as "Indraughts", a term applied to the action of tidal currents in bights and bays along the coast. "Indraught" applying to the set of the flood tide, "outset" to the ebb.

Uschante—The Island of Ushant, or Ouessant, 10 miles off the N.W. coast of France, in the department of Finisterre. The shores are steep, craggy, and surrounded by rocks.

Use—Ile d'Yeu bears from the Ile du Pilier S.S.W. ¼ W., and is 19 miles distant from the Four lighthouse S. ½ W., 37 miles, and from Belle Ile S. by E. ½ E. distant 45 miles. The island is 5 miles long and 2 miles broad, and has an extent of 77 square miles. The town of Port Breton and a fort are on its northern side.

Vaion—*See* **Baion.**

Vamborough—Bambrough Castle, on the coast of Northumberland, standing on a rocky foundation of considerable elevation.

> "Thy tower, proud Bamborough, marked they there,
> King Ida's castle, huge and square,
> From its tall rock look grimly down,
> And on the swelling ocean frown."
>
> (*Marmion*, Canto ii.)

Vas Glenaunt—Iles de Glenant, or the Glenan Ilands, a cluster of small islands, surrounded by rocks both above and under water, some extending 2½ miles from the main body. The navigation of these islands is beset by dangers, and the warning in our text amply justified. The Iles de Glenant, also known as the East Penmark Islands (*q. v.*), are off the coast of Finisterre.

Velafade, Toure of—The old Head of Kinsale was also known as Cape de Velho. On it stood a ruined castle with three towers, the centremost of these being a good landmark. A lighthouse, seen for a distance of 23 nautical miles at sea, now stands here. The bearings, however (north and south with Waterford), given in the text are incorrect.

Wasshe groundis, The—Watchett, on the south shore of the Bristol Channel. The approach to this place is obstructed by a shaft of rocks and beds of rolling stones.

Watir forde, Toure of—The high, white tower east of Waterford haven, since replaced by the Hook lighthouse, visible at sea for a great distance.

Waymouth—Weymouth.

Welbank—The Well, a large shoal south of the Dogger bank.

Wiet—Isle of Wight, anciently called Wiht.

Weris—The weirs or dams raised to protect Harwich harbour from the sea.

Whitevies half—Whitby Haven. The haven is almost dry at low water.

Wiklowe—Wicklow.

Winterbornes and **Wyntir burnes**—Winterton, on the coast of Norfolk, north of Yarmouth.

Wolueshorde—On some old charts marked "Wolveshord", on others Wolfert's Head, at the southern extremity of the Isle of Wight, now St. Catherine's Point. The old name may still be traced in Woolverton, a ruined gabled manor house, said to have been built by John de Wolvert in the reign of Edward I. The *Safeguard of Sailers* (p. 41) calls the headland "Wolfer horne".

Wose and **Woyse**—Ooze or mud.

Yokelis, The—Youghal, south coast of Ireland, on old charts written Yoghill.

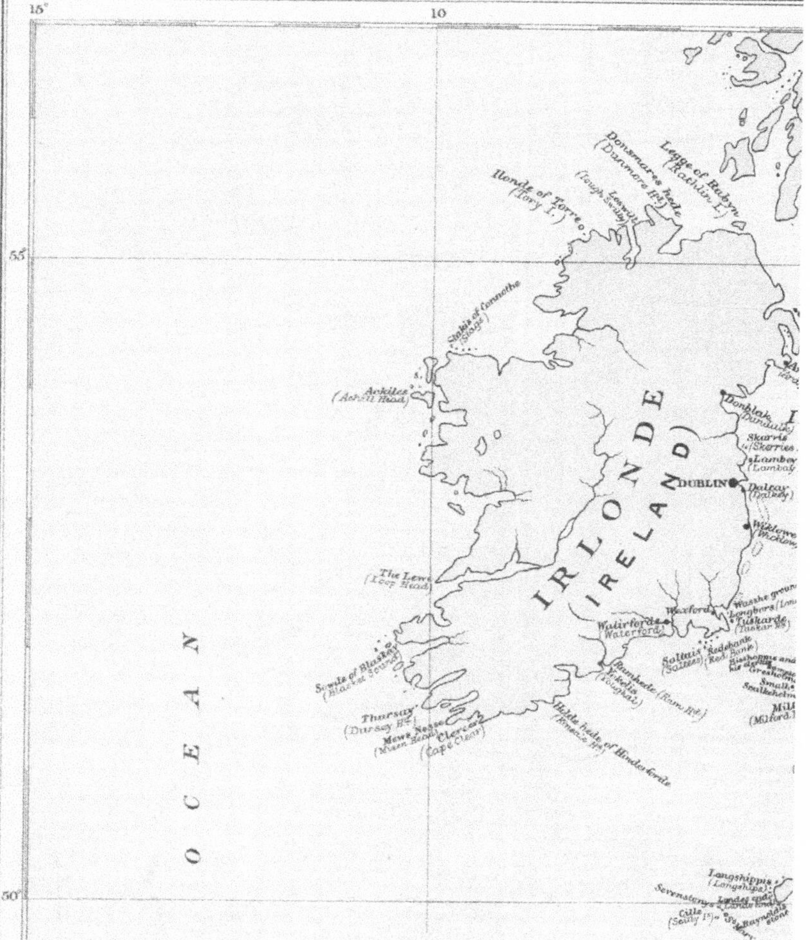

CHART TO ILLUSTRATE

SAILING DIRECTIONS FOR THE CIRCUMNAVIGATION

OF ENGLAND

AND FOR A

VOYAGE TO GIBRALTAR

(FROM A 15TH CENTURY M.S.)

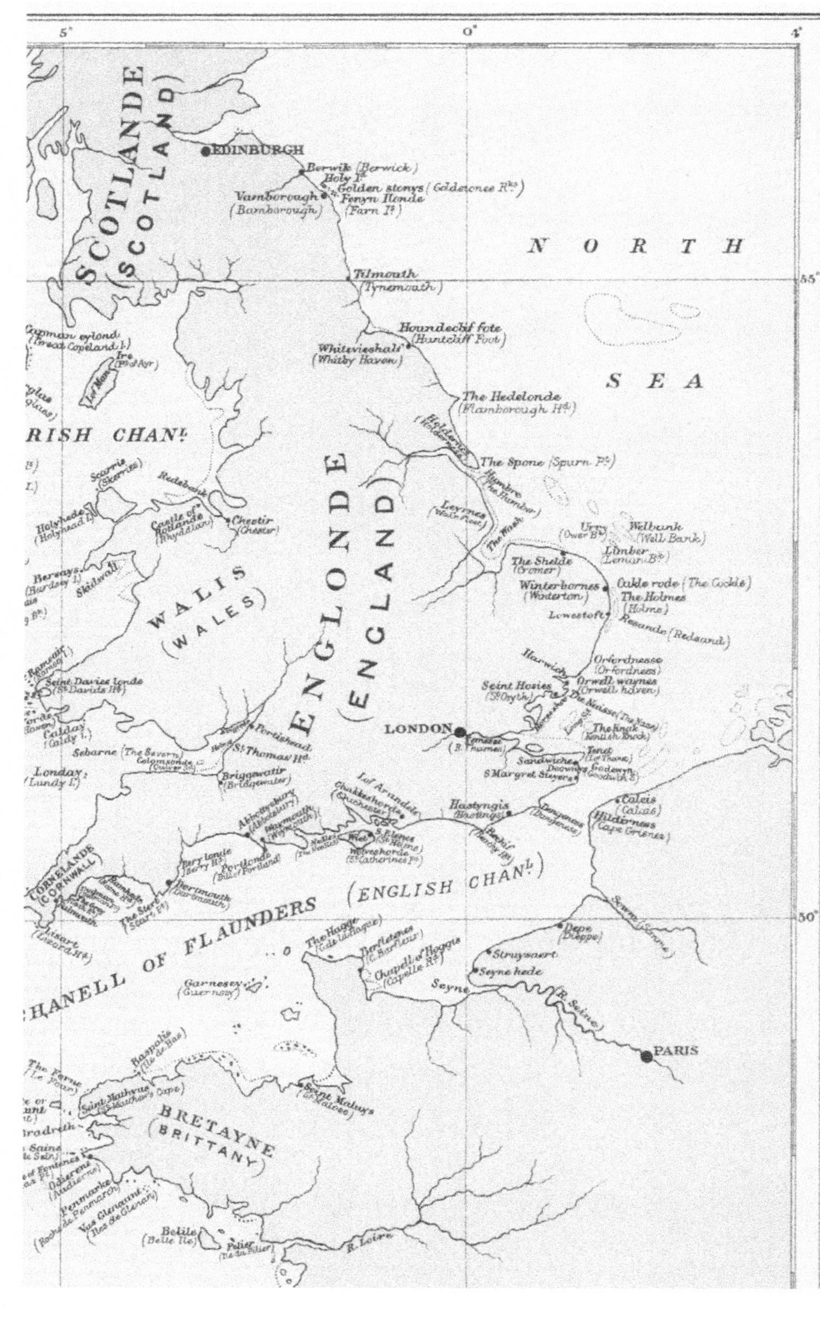

5°
0°
4°
55°
50°
SCOTLANDE
(SCOTLAND)
EDINBURGH
Berwik (Berwick)
Golden stonys
Vamborough
(Bamborough)
Tilmouth
(Tynemouth)
Houndeclif fote
(Huntcliff Foot)
Whitevieshalf
(Whitby Haven)
N O R T H
S E A
The Hedelonde
(Flamborough Hd.)
IRISH CHANL.
The Spone (Spurn Pt.)
Scarris
(Skerries)
Holyhede
(Holyhead I.)
Castle of Rotlande
(Rhyddlan)
Chestir
(Chester)
ENGLONDE
(ENGLAND)
Welbank
(Well Bank)
The Shelde
(Cromer)
Winterbornes
(Winterton)
Lowestoft
Resande (Redsand)
WALIS
(WALES)
Orfordnesse
(Orfordness)
Harwich
Saint Hosies
Orwell waynes
(Orwell haven)
Seint Davies londe
LONDON
Portishead
St. Thomas' Hd.
Sebarne (The Severn)
Londay
(Lundy I.)
Briggewatir
(Bridgewater)
Sandwiche
Caleis
(Calais)
Hastyngis
(Hastings)
Dengenesse
(Dungeness)
Hildernesse
(Cape Grisnez)
Abbotisbury
Weymouth
Portlonde
(Bill of Portland)
CORNELANDE
(CORNWALL)
Dertmouth
(Dartmouth)
The Stert
(Start Pt.)
(ENGLISH CHANL.)
CHANELL OF FLAUNDERS
The Hagge
Depe
(Dieppe)
Struysaert
Seyne hede
Seyne
R. Seine
PARIS
Garnesey
(Guernsey)
Baspolis
(Ile de Bas)
The Ferne
Saint Mathews
Saint Maluys
BRETAYNE
(BRITTANY)
Odierene
Penmarke
Belile
(Belle Ile)
R. Loire

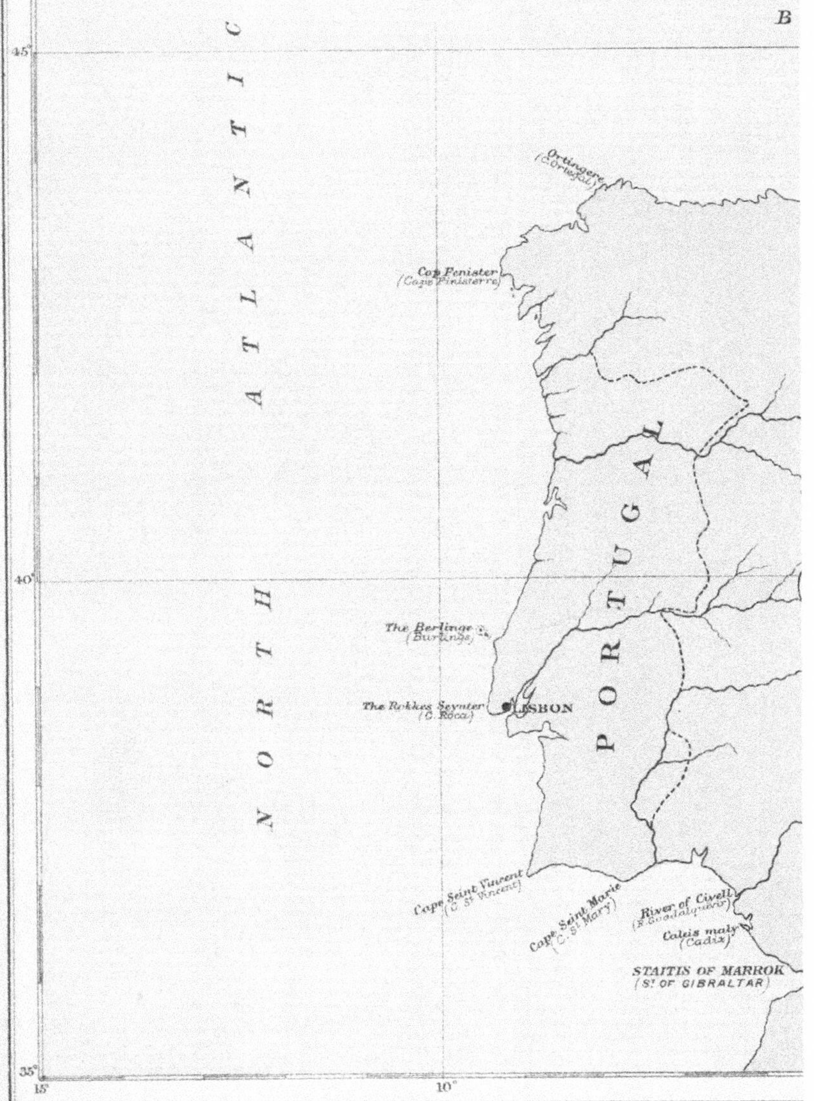
Drawn for the Hakluyt Society
by Mr E. A. Reeves of the
Map Room R.G.S.
B
45°
NORTH ATLANTIC
Ortingare
(C. Ortegal)
Cap Fenister
(Cape Finisterre)
PORTUGAL
40°
The Berlinge
(Burlings)
The Rokkes Seynter
(C. Roca)
LISBON
Cape Saint Vincent
(C. St Vincent)
Cape Saint Marie
(C. St Mary)
River of Civell
(R. Guadalquivir)
Caleis maly
(Cadiz)
STAITIS OF MARROK
(St OF GIBRALTAR)
35°
15°
10°

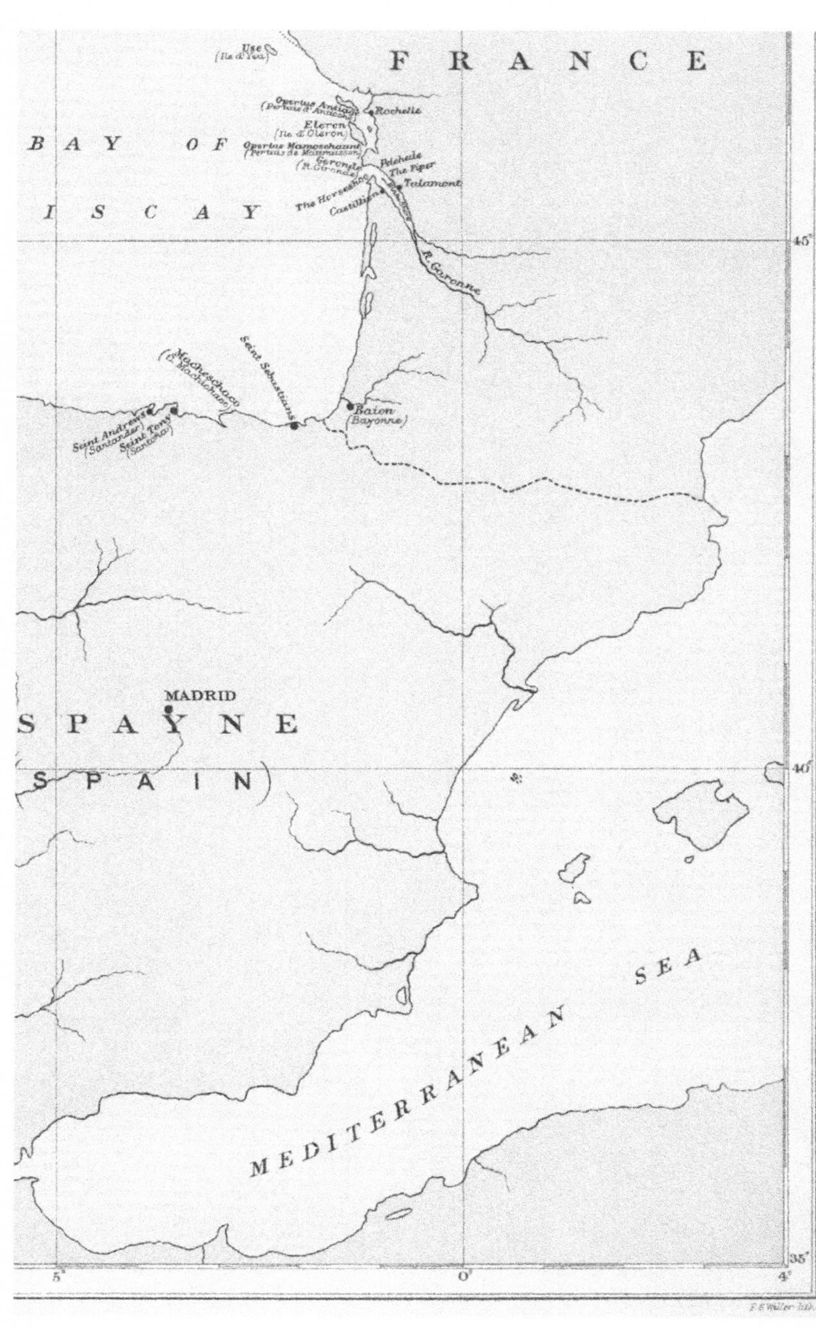
FRANCE
Use
Rochelle
Eleron
(Ile d'Oleron)
Garonde
Talamont
The Piper
Castillion
R. Garonne
BAY OF BISCAY
Saint Sebastians
Baion
(Bayonne)
Saint Andrews
(Santander)
MADRID
SPAYNE
(SPAIN)
MEDITERRANEAN SEA
45°
40°
35°
5°
0°
4°